Cosmic Study

Global Human Mars System Missions Exploration

Goals, Requirements and Technologies

January 2016

SG 3.16
Chair: Giancarlo Genta
Co-chair: Alain Dupas
Secretary: Jean-Marc Salotti

The International Academy of Astronautics (IAA), an independent nongovernmental organization recognized by the United Nations, was founded in 1960. The purposes of the IAA are to foster the development of astronautics for peaceful purposes, to recognize individuals who have distinguished themselves in areas related to astronautics, and to provide a program through which the membership can contribute to international endeavors and cooperation in the advancement of aerospace activities.

Global Human Mars System Missions Exploration
Goals, Requirements and Technologies

Edited by: Giancarlo Genta, Jean-Marc Salotti

Printing of this Study was sponsored by: The Virginia Edition Publishing Company

International Academy of Astronautics
6 rue Galilée, BP 1268-16
75766 Paris Cedex 16, France

Code ISBN/EAN IAA: 978-2-917761-43-4

9 782917 761434

Cover: NASA

Global Human Mars System Missions Exploration Goals, Requirements and Technologies

Editors:

Giancarlo Genta
Jean-Marc Salotti

ACKNOWLEDGEMENTS

This study was conducted under the auspices of Commission III (Space Technology & System Development) of the International Academy of Astronautics (IAA) and benefited from review and comments by numerous members of the Academy, as well as members of the Commission. There were 45 participants in the study with many countries represented, such as: Canada, China, France, Germany, Ireland, Italy, Japan, Russia, Spain, and USA.

The study could not have been completed to this level of detail without the timely and invaluable efforts of a diverse collection of experts from around the world who contributed not only their time and knowledge, but also provided material as well as their technical expertise for the study. These experts included the members of the study group, the chapter authors to whom significant recognition should go, and especially the editors of the study (all names are provided in Appendix A). The diverse contributors included individuals (as well as former staff now retired) from space agencies, commercial firms, universities and non-profit organizations.

Thanks are also due to the members of the International Academy of Astronautics review committees who provided highly useful comments on the final draft of this study report. In addition, thanks to Dr. Giuseppe Reibaldi for supporting the study from inside the Academy and to Dr. Jean-Michel Contant for ensuring that the study results were presented during the 2014 Head of Space Agencies Summit.

Thanks are due also to Emily Czerniewski, Technical Editor at NASA MSFC for revising the manuscript.

The Editors

TABLE OF CONTENTS

LIST OF ACRONYMS

ACR	Anomalous Cosmic Rays
ALARA	As Low As Reasonably Achievable
AM	Additive Manufacturing
ARM	Asteroid Redirect Mission
BLEO	Beyond Low Earth Orbit
COPUOS	COmmittee on the Peaceful Uses of Outer Space
COSPAR	COmmittee on SPAce Research
CSTS	Crew Space Transportation System
DOE	Department Of Energy
DRA	Design Reference Architecture
DRM	Design Reference Mission
DSN	Deep Space Network
ECLSS	Environmental Control and Life Support System
EDL	Entry Descent and Landing
ERV	Earth Return Vehicle
ESTRACK	European Space TRACKing network
EVA	ExtraVehicular Activity
GCR	Galactic Cosmic Rays
GES	Global Exploration Strategy
GER	Global Exploration Roadmap
HIAD	Hypersonic Inflatable Atmospheric Decelerator
HMI	Human-Machine Interface
HMMFI	Human Mars Mission Feasibility Index
HSTI	Human Space Technology Initiative
IAA	Intenational Academy of Astronautics
IAF	Intenational Astronautical Federation
ICAMSR	International Committee Against Mars Sample Return
ICME	Interplanetary Coronal Mass Ejections
IMF	Interplanetary Magnetic Field
IMEF	International Mars Exploration Forum
IMLEO	Initial Mass in Low Earth Orbit
ISECG	International Space Exploration Coordination Group
ISRU	In-Situ Resource Utilization
ISPP	In Situ Proprellant Production

ISP	Specific impulse
ISS	International Space Station
LEO	Low Earth Orbit
LOC	Loss Of Crew
LRV	Lunar Roving Vehicle
LOM	Loss Of Mission
MAV	Mars Ascent Vehicle
MO	Mars Orbit
NEP	Nuclear Electric Propulsion
NTP	Nuclear Thermal Propulsion
PEL	Permissible Exposure Limit
POF	Probability Of Failure
PVA	PhotoVoltaic Arrays
REID	Risk of Exposure Induced Death
RWGS	Reverse Water Gas Shift
SAA	South Atlantic Anomaly
SEP	Solar Electric Propulsion
SLS	Space Launch System (NASA heavy launcher)
STEM	Science, Technology, Engineering and Mathematics
STP	Standard Temperature and Pressure
TEI	Trans-Earth Injection
TMI	Trans-Mars Injection
TPS	Thermal Protection System
TRL	Technology Readiness Level

EXECUTIVE SUMMARY

The idea that humans could reach Mars is quite old and was advocated by many pioneers of spaceflight. Apart from fictional descriptions, sometimes bypassing completely the problem of getting there, and of pioneristic work dealing with the general aspect of the problem, the first detailed study of a human Mars mission was done by Wernher von Braun who published in 1949 *Das Mars Projekt*, a technically sound project, demonstrating that it was possible to reach Mars with a technology predictable for a not too distant future. This project, although technologically consistent, didn't take in due account the relevant costs and, as perhaps unavoidably in a first attempt to rationalize the problem, was not sustainable.

Since then many projects were published by space agencies, individual researchers and companies from many different countries, but after 65 years human Mars exploration seems to be still a goal far in the future. One of these studies was the "Cosmic Study" on human Mars Exploration published by the International Academy of Astronautics (IAA) in 1993.

The International Space Exploration Coordination Group (ISECG) produced a Global Exploration Roadmap (GER) in which three destinations for exploration missions are contemplated: the Moon, the cislunar space and the asteroids, and Mars. While in the roadmap the goal of human Mars exploration is considered as a goal to be achieved in a more distant future, several "stepping stones" are stated, among which there is a strong program of Mars robotic exploration.

In this context, the IAA in 2012 has set up a Study Group (SG 3.16) to produce a new Cosmic Study dealing with human Mars missions. The first results of this work were summarized in a White Cosmic Study presented at the Heads of Space Agencies meeting in Washington D.C., USA in January 2014. The final results of the study group constitute the present document.

The purpose of this IAA study is not to produce yet another project design, but rather to help future projects develop by summarizing, in one document, the current state of the art on the various aspects related to Human Mars Exploration: technological, human, economic and organizational. It will end with specific recommendations for overcoming the obstacles still cluttering the road toward achieving the end goal of launching a manned mission to Mars.

The starting point of the study is that the very ambitious goal of bringing humans to the Red Planet must be pursued as a truly international enterprise; in which all countries contribute toward the final achievement. Another starting assumption is that reaching Mars must not be a single exploit, but a very important step along a road leading the human species to become a spacefaring civilization.

The study is subdivided into 10 sections.

1. **Mission rationale.** There is no doubt humans have an innate drive to explore; however, we cannot just say that we want to go to Mars *because it's there*. There is no way we can justify a difficult, costly and risky enterprise like a human Mars mission in such a simple and naive way. Justifying a human mission to Mars in the near future is a tricky business, and several hundred pages have been written on the subject without reaching a definite conclusion.
   This section tries to compound pragmatic with aspirational rationales and to weigh them against the technological feasibility, the evaluation of the risks, and the affordability of a mission of this kind. Several roadmaps have been proposed, which realistically assume that several steps must be performed before attempting such an important enterprise.

2. **Lessons learned from the past projects for Human Mars Exploration.** This section is not an attempt at making a detailed history of the various projects for human Mars exploration that have been devised over the last 60 years, something which would require several volumes. Instead, it presents a brief overview of the basic ideas underlying the various mission concepts and systematically describes the main causes of mission failure. Some lessons learned from these past projects are highlighted, with particular stress on the affordability of the various approaches and the reasons behind their presumed failure.

3. **International Cooperation for the Human Exploration of Mars.** As already stated, the basic assumption underlying the present study is that Human Mars Exploration is a common enterprise of humankind and must be a cooperative effort between nations, space agencies, industries and non-governmental organizations. This chapter will elaborate on this subject, starting from the present international roadmaps like the ISECG-GER. The Human Mars Mission studied here is to be seen as a sort of continuation of the cooperation seen in the International Space Station (ISS) on a larger scale.

4. **The environment.** The environment humans will have to face when on Mars is a harsh one – very low atmospheric pressure, freezing temperatures, high

radiation, absence of life (or, at least, absence of widespread life), etc. The environments the astronauts will have to cross during their journey to and from their destination is even harsher. This chapter summarizes the main characteristics of these environments, with a special focus on the planet's surface and satellites._The problems of planetary cross-contamination and protection are also delt with, especially with regard to the most important problem: the possible existence of alien life, either past or present.

5. **The human issues.** Human issues are most important in all human space missions, and in particular during long and difficult ones like a voyage to Mars. The biggest physiological problems humans will have to face on their way to the Red Planet, and even when on its surface, are those due to low gravity and to radiation. But physiological problems are not the only ones, since such long missions, so far from home, are a potential source of cognitive and psychological problems. Such problems potentially have a strong impact on the mission and choices such as the number of participants can strongly influence the outcome of the mission.

6. **The space transportation system.** One of the main issues related to a Human Mars Mission is how to get there and how to get back. There is a consensus on the long stay option, the use of in situ resources to manufacture part of the propellant to get back to orbit and the pre-positioning of assets on the surface of Mars before the launch of the crewed vehicle. There are, however, different opinions about the choice of the interplanetary transportation systems for the first missions. One group suggests the simplification of several important parameters of the mission to make it affordable and achievable in the near future: three or four astronauts, chemical propulsion, aerocapture for all vehicles, junction in Mars orbit rather than in Low Earth Orbit (LEO) and no more than four heavy launchers (130 tons LEO capability) to send everything to Mars. Another group suggests the development of nuclear based propulsion systems–nuclear thermal or nuclear electric–to reduce the amount of propellant and pave the way for game-changing technologies. These issues are still much debated, since using well consolidated technologies reduces some costs, development time and risks, but developing new technologies may reduce other costs and improve performances, a thing which in turn may reduce other risks and costs.

7. **The planetary infrastructure and vehicles.** Once on Mars, the explorers need a place to live, devices for surface exploration, a power plant, In Situ Resource Utilization systems to produce as many commodities as possible from local

sources, and much more. Alternative solutions have been proposed, and a discussion about them is required. The design of habitats must take into account not only the technical requirements but also the physiological and psychological needs of the people who will live in them. Apart from the habitat, this section deals with the power system, the In Situ Resources Utilization (ISRU) systems, the exploration vehicles and other required equipment. The issue of planetary protection is an important conditioning factor in this matter.

8. **The ground sector.** Although apparently a less important issue, a number of infrastructures on Earth are instrumental to mount a Human Mars Mission. They include the communication network, the ground control centers, the astronaut training facilities and the various labs for performing ground simulations of the necessary devices. Since the mission will be performed by a number of nations, the centers shall be distributed among the various participants, which may cause political problems.

9. **Mission architecture options and roadmap.** The various alternatives lead to different possible architectures, and the goal of this chapter is to describe two of them. Moreover, the road leading to a human Mars mission is made of a number of preparatory missions. Two specific mission concepts are highlighted: The first is a heavy robotic Mars sample return mission, which could make use of a heavy launcher, test interplanetary propulsion systems, test the entry, descent and landing of heavy payloads on Mars, and test in ISRU systems. The second is not exactly defined but would primarily test and qualify the habitat for a three year journey without resupply. If chemical propulsion is chosen for interplanetary transportation and the mission is simplified, it is suggested that the first human mission to Mars could be undertaken at the beginning of the 2030. Otherwise, the roadmap should include the test and qualification of advanced propulsion systems, and the first mission could occur the following decade.

10. **Conclusions.** Some conclusions of the stydy are summarized in this section.

11. **Recommendations** The study concludes with a number of recommendations whose aim is to facilitate the implementation of human Mars missions. The recommendations deal with the development of the technologies which are required to implement a human Mars mission, but above all deal with the human, political and economical aspects of the matter. These reccomendations stress the need for a roadmap that includes a number of missions which can be

used as stepping stones toward the goal of not only sending humans to Mars, but bringing them home safely.

References and Appendices. A list of references, subdivided by chapter is included. The appendices report a list of the members of the Study group, a list of the mission architecture options, some considerations about the propulsion systems, and a proposal to introduce a Human Mars Mission Feasibility Index.

Chapter 1
MISSION RATIONALE

1.1. Scope and goal of the study

Since the swift termination of the *Apollo* program in 1972, the question of the continuation of space exploration, and in particular for a human Mars mission, has persisted on the agenda of space agencies and, on several occasions, even led to political initiatives to restart the effort. During those four decades, a profusion of projects were studied and proposed. While none of them went past the report stage, this lengthy period allowed the ideas about the mission architecture to mature clearing out some early misconceptions (e.g.: preference for a short stay opposition scheme, view of the Mars surface as a more dangerous place than space itself), and allowing the emergence of efficient technical innovations (e.g.: aerocapture at Mars, In Situ Propellant Production). Nevertheless, those proposals suffered, to different degrees, from some weaknesses:

- They were mainly engineering studies, with little consideration given to criteria other than operational capabilities, technical performance and (less often) overall cost. More specifically, they dealt mainly with mission and hardware design, much less with the overall project planning; yet, it is this aspect of any major techno-scientific project which is most significant for the decision process.
- Most proposals were built on the basis of preset science objectives, without trade-offs between technical objectives and needed financial resources; for instance, the sequence of NASA architectures (Design Reference Mission (DRM) one through five and Design Reference Architecture (DRA)) did not consider various crew sizes, which have a significant influence on the project cost.

This situation is at the basis of the motivation for this study. ***The intent of this IAA study group is not to produce yet another project design, but rather to help future projects build-up*** in:

- Surveying the spectrum of criteria which such a project should satisfy in order to
  - Be approved
  - Remain continuously supported

  - o Be finally successful, that is achieving the goals under the imposed constraints.
- Analyzing the objectives of each mission phase and their requirements.
- Examining the many options which are available and establish the rationales of choices, taking into account their influence on the objectives.

With that in mind, it should be possible to recognize preferable options, and to build a hopefully attractive and feasible project proposal. ***Attractive and feasible***, are both critical terms.

In the past Mars mission studies, usually the project was considered successful if a high mission success probability was credibly established. But although decisive, making this goal the key priority carries the serious risk of missing the indispensable preliminary objective: that of making the project so attractive, safe and, feasible enough to have a chance to be funded and get started!

More precisely, this objective means:

- That the project technical and financial feasibility is properly established and, even more decisively,
- That the many associated risks are worthy of the proclaimed gains. However smart and worthy a project is, as long as it is not backed by thorough arguments about its significance for the funding partner's policies, the whole thing is doomed to either remain at the status of proposal or collapse.

The consequence is that in designing such an overarching project, one must be very attentive in meeting the criteria required for securing the decision to start the project and the continuous commitment of the partners. Before feasibility (cf. §I.3), *attractiveness* needs to be properly established, balancing

- ***Foreseen returns*** on investment, in terms of knowledge acquisition, industrial innovation and economic development (including employment), geostrategy linked to international cooperation, societal impact (education, population morale, national pride) and others;
- **Known *risks*** which, while associated with many different domains, are–in the end–political: loss of crew, loss of mission, program costs overrunning, delays, oversold scientific returns, technical dead-ends, discontinued support from partners, unmanageable international cooperation, and many others. Taking appropriately into account the full spectrum of those risks is pivotal for the acceptance and the success of the project.

These aspects relating to the appeal of the Human Mars System Exploration project, not sufficiently covered in most of past project studies, have been given special attention is this report, in fact the same attention of technical (and human) feasibility aspects.

1.2. Criteria for an attractive program

A *sound* proposal should aim at optimizing the project technical and operational design, as well as the program management principles and organization. An *attractive* proposal must contain forceful political motivations, with risks convincingly evaluated as acceptable. Motivations and risks are discussed below.

1.2.1. Political motivations

Scientific knowledge acquisition

While it is acknowledged that the scientific goals of Martian exploration could not by themselves justify the sharp increase of resources required by a switch from science obtained through robotic missions to science obtained through both robotic and human exploration, this remains the driver of the program goals.

In this context the scientific community is the prime partner to satisfy. It is thus of the utmost importance to build a project in partnership with scientists and centered on science objectives. The scientists involved should understand that the program will be a strong growth factor for their activity. Moreover, it must be clearly stated that the human presence will not replace robots but, on the contrary, will enhance robotic productivity. Having the science community agree with and back the project is an important prerequisite.

Among the various aspects of scientific research which can be carried out on Mars, astrobiological studies have a particular role, both for their importance and for the difficulty of performing them with robotic devices. On the other hand, the utmost care must be exerted to prvent contamination due to the presence of humans on the planet.

Another key issue is the pacing of the program. Sequential robotic missions should pave the way for a human landing and the value of interim human operations to other destinations should be scrutinized as well as their cost and consistency with the Mars objective.

Contribution to innovation and economic development

It is recognized that such a big techno-scientific program, venturing into the unexplored and to the limits of our knowledge, constitutes one of the most powerful tools in the hands of governments to push forward their nation's innovation capabilities and economy. As it involves industrial strategies, high-tech, and expensive facilities, Mars project activities may generate highly paid jobs which cannot be easily outsourced.

Although most of the required technologies have reached a reasonably high Technology Readiness Level (TRL), pushing them to the limits imposed by unforgiving requirements, combining them into the most complex space system ever conceived and to which human lives will be entrusted, is a formidable challenge. The boundaries of aerospace industry's and research organizations' ingenuity and manufacturing skills will all be pushed forward, leading to increased quality and efficiency, new and better products and market share gains. The forceful impact of the program on economic development should be documented through assessment of a number of indicators, mainly:

- Identification of domains where progress will occur;
- Forecasts of industrial innovation and new application areas;
- Workforce enrolled.

Finally, an important contribution to economic development stays in motivating young generations to get higher education and enter STEM (Science, Technology, Engineering and Mathematics) careers, particularly in countries lacking young specialists in these areas.

Geostrategy and international cooperation considerations

With the outstanding development of its commercial applications as well as in the fields of resource and environmental monitoring, science, security and defense, space has become a strategic arena for developed and rapidly developing nations. Those nations willing to maintain or reinforce their international influence cannot neglect any of the main sectors of this new domain of activity. So much so, in fact, that when one or more of the leading space powers moves in a new direction, others are incited to follow.

When this happens in an area of space exploration perceived as an endeavor in the name of all humankind and a tool for peace and global development, this is even more likely. It is widely admitted that, even though started by a single nation, the Mars mission program will become an international collaboration. This is a basic assumption in this study.

Program planning and international cooperation should satisfy several criteria relating to efficiency and to long-term robustness:

- The program structure and conceptual organization should be designed with special care to facilitate sharing work and responsibilities, while preserving overall efficiency.
- International partners should be able to satisfy their national interests, which may be many and very specific: some will want to participate in order to strengthen their leadership, others to upgrade their rank, others to acquire technologies, others to reinforce their diplomatic links, and so on.
- Presenting the program as an effort toward peace and global development should be explicitly said in defining project objectives and public communication policy.

Contribution to societal aspects

As globalization leads to accelerated diffusion of knowledge and activities, the strategic character of innovation and of entrepreneurship is recognized more than ever. Even if not every concerned nation is yet up to the challenge, each knows that a sufficient level of Research & Development effort is necessary to defend its economy and, in the end, its wealth. However, direct financial commitment is far from being sufficient to realize a human Mars mission. What is also required is attracting an appropriate flux of students to STEM careers, to augment the creative workforce that will innovate technology and create new business.

Also it is necessary to have the public understand and accept that a fraction of their taxes will have to support such long-term objectives. This is not easy and requires significant confidence in the future as well as a clear demonstration of future benefits

From these two points of view, starting a comprehensive Mars exploration program, offering it to the public, and more specifically to the youth, the perspectives of exciting discoveries, of new activities, and a reason to dream, would certainly be really a smart move. Thus, when laying out the program, much attention should be given to factors enhancing:

- The level of public acceptance / support in major spacefaring nations and other participating countries.
- The interest of youth, and more specifically the involvement of students to play a role in the project through their Colleges and Universities.

Long terms goals

Although a controversial aspect, many hold that the true motivation of human Mars missions is starting the road which eventually will transform humankind into a spacefaring civilization, with bases on several planets. In this vision, Mars would be just the first planet to be colonized and settled by humans, possibly after having established an outpost or even a colony on the Moon.

In this vision, long term goals are the only ones ensuring the funding continuity required for the exploration effort. Continuous support will prevent the same outcome as the *Apollo* program, and the main surce of funding continuity is linked with the exploitation of Mars for economic purposes, which is of primary importance in granting the sustainability of the whole effort.

Some long-term goals include giving humankind a safe haven in case major disasters struck Earth, for instance an asteroid strike or another planetary catastrophe. These goals can hardly justify a Human Mars Mission, but can act to reinforce other motivations. The consensus is that no single motivation can solely justify such a complex, costly, and risky enterprise: this can be justified only by the convergence of several motivations.

1.2.2. Risks: political acceptance

Besides the technical and financial feasibility of the project, mission risks also play a major role in the decision making process. The technical and financial experts' input, reviews and, credibility can go a long way toward determining overall project feasibility. However, evaluating risks, giving confidence to evaluations and, last but not least, eliciting final mission approval is much more difficult. This probably explains why, however decisive it is, this aspect of the project is so rarely discussed.

The risks involved are both numerous and varied, thus complicating the tradeoffs.some risks have different significance to different actors; for instance, the risk of loss of crew is probably even less acceptable to policymakers, due to its programmatic and political consequences, than to the mission astronauts themselves.

Risks ranging from strictly technical to the policy makers endorsing the program pervade the whole program structure through the hierarchical chain of responsibilities and through the causal chain of consequences of a blunder. However, as far as acceptability is concerned, it is ***the information and perception by top decision makers*** that are key points. Thus the project should be planned focusing on:

- Clearly describing what is important for decision making, in terms of gains and risks, and in the way charts and comments are presented.
- The frank description of existing differences in evaluating risks.
- The relevance and impact of risks to overall program.
- The level of uncertainty.
- The way to balance risks and issues of much different nature (e.g.: how to balance a reduction of the probability of Loss of Crew with a development cost increase?). What follows is a tentative and not exhaustive list of risks to be documented and discussed:

- **Costs overruns**, concerning:
  - Total development cost.
  - Annual development budget load.
  - Annual operational cost.

  Resulting from:
  - Improper technical challenge evaluation.
  - Over evaluation of a team member capability or commitment.
  - Poor overall management.
- **Delays,** concerning:
  - Development.
  - Operational missions tempo.

Resulting from:

- Same causes as above.
- Lack of mission reliability.

- **Program slowing-down or termination**, resulting from:
  - Overestimated descaling or termination possibilities.
  - Underestimated resulting damages (financial, industrial, and political).
  - Underestimating cost and difficulty of objectives.
- **Loss of Mission** (LOM), resulting from:
  - Poor LOM probability evaluation.
  - Excessive self confidence.
- **Loss of Crew** (LOC), resulting from:
  - Poor LOC probability evaluation.
  - Excessive self confidence.
  - Inappropriate or biased comparison with past space crewed programs.
- **Lack of program resilience**
  - To a LOM.
  - To a LOC.
  - To partner withdrawal or failure.

It is suggested that a panel of experts from space agencies be formed to produce a comprehensive document on this issue.

1.3. Criteria for feasibility

The feasibility of the project, i.e. the probability that the program operational and safety objectives are attainable with a precisely identified set of technologies, within a predefined envelope of financial resources and in a specified time, is usually the main driver of the project studies.

1.3.1. Technical feasibility

The two most significant criteria involved when evaluating the technical feasibility of the project: the ***technologies readiness level*** (TRL) and the ***complexity*** of the system (number of modules, equipment, critical operations, etc.).

In many cases, but not always, it is wiser to ***prefer a high-TRL technology*** (flight-proven when available) rather than one with higher performance potential but low TRL. This strategy reinforces credibility and typically reduces development cost and risk. Even though doing so may increase mission cost, this strategy facilitates technical evaluation and decisions. Note that using Hohmann trajectories mission frequency cannot be more than one every 26 months. The argument in favor of this strategy is that the more one relies on proven technologies and data the easier it is to establish a robust technical design file giving confidence in the engineering.

However, sometimes better but unproven technologies could be developed before a Mars mission starts. Moreover, a mission to Mars appears even now to need at least some low TRL technologies, for instance In Situ Resource Utilization (ISRU) or aerobraking of large entry vehicles.

A typical example is propulsion: while chemical propulsion is probably barely capable to enable a human Mars mission, the advantages of Nuclear Thermal Propulsion (NTP) or of the lower TRL Nuclear Electric Propulsion (NEP) are such that many think it is worthwhile to rely on it (see Sections 6 and 9). These are still open issues, depending on many factors that are either unclear or unknown at this time.

The choice between a proven but less advanced technology and one more advanced but with a lower TRL depends on:

- The expected timeframe of the mission (if the mission is expected to take place in a distant future, it is possible to bring at the required TRL more advanced technologies).
- The overall roadmap for human exploration and he possibility of introducing technology validating missions into this roadmap.
- The number of missions planned (the investment to develop new technologies must be compared with the savings the new technology will bring to each mission. Above a certain number of missions the new technology will save costs.
- The trend in space activity, including private commercial missions.
- The results of R&D in connected areas. For instance, a better evaluation of the dangers due to radiation can increase or decrease the importance of a reduced time in space and so can increase or decrease the importance of developing advanced propulsion.
- The still not quantifiable opportunities to innovate offered by the Mars program.

1.3.2. Achieving scientific objectives

Realistic scientific objectives of the Mars mission must be consistent with the technical means used. The demands of the science community are many and often well in advance of what is affordable in terms of resources (mass, energy, cost) or development risks. Consequently ***science community frustration or misunderstandings between scientists and engineers should be avoided*** for the success of the project.

Many of the options described in the following sections have a significant impact on the attainability of scientific objectives; examples are the choice between short or long stay missions, crew size, determining the mix of different disciplines.

The technical description of the project should be sufficiently detailed with regard to what concerns the scientific equipment and the operational plan, in order to ***demonstrate that the project science agenda is sound and agreed upon by the scientific community***.

1.3.3. Affordability

Affordability implies

- Gathering costs from in-depth industrial studies led by space agencies or governmental cost centers.
- That existing studies should reference past programs, in terms of predicted and actual costs, costing methods and data bases.
- That they should account for technical, scientific and financial risks.
- That the risks posed by unexpected costs and delays are at acceptable levels and consistent with the scope of the project.

Economical affordability must be carefully assessed. That space exploration, including human Mars missions, is expensive, but not as expensive as what is spent for other activities, must be clearly stated. In 2012, for instance the total expenditures of all space agencies amounted to 42.0 billion US$[1], (17.9 billion US$ being spent by NASA), which amount to just to 2.6% of the total military expenditures in the same year (1,617.7 billion US$[2], 739.3 by the USA) and are comparable with what is yearly spent in other large infrastructure projects.

Increasing current space budgets is probably required to allow a human Mars mission, but expenditures will be spread over several years. The NASA budget, which after 2000 has been consistently between 0.5% and 1% of the US federal budget, was at 4.5% at the height of the Apollo Program[3]. Human Mars exploration, particularly if conducted as a true international effort, will affect the budgets of participating countries much less than what the *Apollo* program costed the USA in the 1960s.

To reach economical affordability it is necessary that the actors involved, and mainly the industrial partners, can make a profit from the investment they make in mounting this mission. In particular, only if the various missions, perhaps except the very first one, are successful in creating economic returns it will be possible to

[1] Organisation for Economic Co-operation and Development, 2012.
[2] Stockholm International Peace Research Institute, 2012.
[3] Source U.S. Office of Management and Budget.

avoid that human Mars missions end up like the Apollo Program but are successful in opening an era which will transform our society into a space faring civilization.

1.3.4. Political feasibility

Political feasibility is linked to several aspects: the possibility that each single participating nation takes the decision of entering this enterprise and the possibility that this deision is actually ratified and maintained for the required time by the governing bodies of that nation.

The first depends also on the international atmosphere prevailing at the moment in which the decion is taken, and may be jeopardized by a worsening of the political climate. While a bad international political climate was a factor which strongly contributed to the beginning of the space age, in which the various enterprises were mostly undertaken by either one of the superpowers, an international enterprise like the Human Mars Mission here studied requires a favourable international situation.

From this viewpoint, an encouraging factor is the fact that the worsening of the Russia-USA relationships in the last years didn't have negative outcomes on the ISS, but probably this shows that a large international venture can survive a worsening of the international situation, but not that the relevant agreements can be started in this condition.

The second point depends largely on the political structure of the various states and on the stability of the various governments. The economical feasibility is certainly important in convincing the decision makers. Since the technological and industrial development even in fields outside the aerospace industry can cause a decrease of the costs, it is predictable that the introduction of new technologies and the increase of the role of privates in space will make things easier.

Another important point is the popular support, which requires a strong outreach activity by the organizations which advocate space exploration and, specifically, human Mars exploration, and an increase of the scientific and technological literacy of the general population.

The general development of commercial activitis in space and the development of private space activities will be a strong incentive for taking the relevant politiocal decisions.

1.4. Main relevant options

During the past decade some mission choices have been settled but the list of those remaining is still long. They involve mission architecture and technical and operational issues.

A sound proposal should aim at optimizing the project as well as the program management principles and organization, with the clear goal of making the human Mars Exploration program both appealing and affordable to the political decision makers. In this context all options, from the most global to the most specific, should be given thorough attention, and the corresponding choices related to its influence on the relevant program attributes and criteria.

Some of the options identified are:

- Conjunction or opposition (short or long stay) mission
- Number of missions and landing sites
- Crew size
- Interplanetary propulsion system
- Mars orbit insertion
- Descent vehicles and EDL strategy
- ISRU options
- Launcher to LEO strategy
- Spacecraft architecture
- Overall redundancy and multiple missions strategy
- Preparatory missions and roadmap.

A detailed discussion of these and other options are in the following Chapters.

An extensive draft list of options, together with their implications, is reported in Appendix B. It can be used as a check-list in discussing the impact of design choices. As in all decision-making, the difficulty is in weighing the impact of usually very different or opposing factors as, for instance, crew safety vs. development cost. This is where engineering and programmatic expertise of space agencies, industrial contractors, program managers, astronauts and others should be requested. Also, to cope with complexity, the project should be founded on a heuristic process based on a hierarchical approach to planning and design. In practice the higher level decisions should be made first, providing guidelines for more detailed ones.

Chapter 2
LESSONS LEARNED FROM PAST HUMAN MARS EXPLORATION PROJECTS

2.1. The pre-space period

In 1949, a technically sound project (*Das Mars Projekt*) [1], issued by a renowned rocket engineer (Wernher von Braun) proposed for the first time how to reach Mars. Although firing the imagination of many, this work did not account directly for practical feasibility or for cost. For this reason, while keeping its historical value, it did not result in an action.

Lesson*: whatever its technical value (and its scientific interest), a project giving no consideration to acceptability and affordability should not be proposed.*

2.2. The post-Apollo withdrawal from Human Space Exploration

In the 1960s several projects for human Mars missions were developed by both the US and the USSR, some based on Nuclear Electric Propulsion (NEP). Among the American projects, those by Stuhlinger in 1957 and 1962 are worth mentioning. In particular the latter was quite detailed and included ion thrusters (Figure 2.1).

Figure 2.1 Nuclear-electric spacecraft designed in 1962 for a human Mars mission.

This spacecraft, designed for a 1980s mission, included details still discussed today, such as rotation to achieve artificial gravity, a radiation shield, a space radiator and others.

Also based on NEP were Russian designs like the Martian Piloted Complex (TMK), whose aim was a Mars flyby, and the Mars Expeditionary Complex which planned a Mars landing. The spacecraft would have been launched by the N1 rocket and all plans were canceled after its two failures.

Other projects were based on Nuclear Thermal Propulsion (NTP), hoping it could be developed in a shorter time, and in fact much efforts were spent in developing NTP in the US and USSR.

The success of *Apollo*, and the maturity obtained in nuclear rocket technology led NASA to propose a human Mars exploration program as the logical follow-on. But the race against the Soviets was won and the level of resource commitment of the *Apollo* era could not be sustained any longer during the Vietnam War.

Nevertheless, in order to preserve the newly acquired industrial and technical space assets, as well as to not appear to withdraw from space, the Shuttle and Space Station programs were eventually approved. In fact, to prove that the space exploration was still important, the Space Station was depicted and marketed as an essential stepping stone to the Moon and Mars, serving as an assembly and refueling station. It took not many years for this seducing picture to fade away.

***Lesson**: proposing a program without putting affordability (consistency with the budgetary situation) and political appeal (consistency with national policies) as main priorities is unproductive.*
***Lesson**: hiding a political withdrawal behind a brilliant tapestry stating that Human Space Exploration is continued while the contrary is true is a convenient practice. However, it should be avoided, as in the long term it weakens public confidence and support.*

2.3. The Space Exploration Initiative (SEI, 1989) and the Mars Direct conceptual approach.

On the occasion of the *Apollo* 20th anniversary, the US President at the time, G.W. Bush, launched the SEI. But the initiative was immediately killed by Congress, as a consequence of the disastrous NASA's "90 days study". This huge project demonstrated that NASA was still in the state of mind typical of the era of unconstrained budgets tied to the Moon race; it didn't put the required stress on affordability, with preference given to beautiful and high-performance technology and to rich and complex system architectures.

As a reaction to this outdated approach the Mars Direct project emerged proposing cost-saving innovations (aerocapture, In-Situ Propellant Production (ISPP), direct return) and and putting some options to rest permanently (long stay, split mission, safe haven on Mars). While early plans called for the development of nuclear propulsion, either thermal or electric, the Mars direct approach was based on chemical propulsion, so that no protracted technology developments were required.

Lesson: *proposals should adapt to the evolving political and budgetary context; official organisms and institutions are prone to lag behind changes.*
Lesson: *Mission designers should:*

- *Be straightforward: every item or effort should serve to achieve the goal efficiently;*
- *Keep the design simple, even at the price of less performance or flexibility.*

2.4. The NASA Design Reference Missions (DRM, 1993 - 2009)

The Mars Direct innovative project was considered optimistic and expensive, but it proved sufficiently appealing to revive the humans to Mars idea. NASA liked it and was happy to design on this basis a new Mars mission. From 1993 to 2009, a series of versions of the so-called Design Reference Missions (DRM) were issued. In this period, the design trend has been in most cases to increase complexity and sophistication, as if, once again, NASA understimated affordability as a constraint. For instance, crew size increased, mission return flight was no longer direct, and larger mass margins were considered.

Lesson: *many aerospace engineers tend to increase "sophistication" of projects, either in pursuit of more performance, or of more safety, or under external pressures. With an eye on financial realities this tendency must be kept to a feasible minimum.*
Lesson: *the DRM effort has been largely an American effort. Mission goals, architecture and schedule have been repeatedly discussed in international committees, but most of the detailed design was performed by NASA and its US partners. This could become a problem when a Mars program is proposed, as it is generally admitted that it should be a cooperative endeavor. Thus, in parallel to design studies, and interacting with them, a study is recommended to define the cooperation framework, its mode of operation, and the resulting constraints on partners and on mission design itself (e.g. crew size).*

2.5. Constellation: the Moon first (2004-2009)

In the US, the *Columbia* accident caused many to question the continuation of human spaceflight. Was it worth the cost and risk? By the end of 2003, this question was settled: The decision was to continue on the condition that a high-level goal was assigned to this endeavor.. This goal was returning to the Moon and, ultimately, landing humans on Mars. The Moon step was presented as a logical stage on the way to Mars, in order to master technologies while complying with financial constraints. Unfortunately, this program was launched with insufficient budgetary resources to be able to meet its objectives.

***Lesson**: launching a program without the adequate financial support may lead to wasting time and money, and loss of public confidence.*

2.6. The Obama era (2009-)

The new president, seemingly, gave human space exploration a lower priority and acknowledged that the related programs were unaffordable. In this situation he canceled the Constellation program. As in the post-*Apollo* era three main arguments were, and still are, put forward to justify to the new Administration policy in an effort to prove that space exploration was still important:

- Adding a return to the Moon step to the program increases the total cost and wastes financial and technical resources; since "we have already been there", the pay-off is insufficient.
- Before committing to a human Mars program, it is necessary to develop new technologies, as we are not ready to go with current ones.
- This technological effort should not be specifically directed toward Mars, in order to free innovation from destination specific constraints.

It is interesting to note that Congress nevertheless decided to continue development of two essential assets of the canceled program, the heavy launcher (now called the Space Launch System, or SLS) and the Orion spacecraft.

***Lesson**: proclaiming lack of preparedness is an even more efficient and convenient means to mask political withdrawal than the advertised role of the Space Station in exploration was; in effect this allows the appetites for technology development actors to be satisfied, while easily tailoring resources to budgetary constraints. The risk is inefficient spreading of financial and human resources with poor or non-existent concrete achievements to show after eight years.*

***Lesson**: without a strong political vision and support, there is no way for a rational, robust and affordable Mars program to materialize.*

2.7. The emergence of private initiative and new players

Since several years ago a number of wealthy private entrepreneurs have entered the space domain. The appearance of SpaceX as a new and major competitor in the field of rocket launches is the most evident landmark. SpaceX has publicly stated its goal to develop a low cost transportation system to LEO and, eventually, to Mars. Its cofounder Elon Musk has said that colonizing the red planet is the goal of his life. Other remarkable initiatives deserve consideration even if they still have to come to fruition, as those dedicated to asteroid mining (Planetary Resources, Deep Space Industries), robotic missions to the Moon, and space tourism (Space Adventures, Bigelow). Conversely, the Mars One initiative appears instead much less realistic and has lost support.

At the same time, while the two original players in the space race of the 1960s went on producing a number of projects for a human Mars mission and launching robotic missions, which were often presented as stepping stones toward the goal of a human landing on Mars, other countries and space agencies have acquired the capability of reaching the Red Planet and manifested their willingness to engage in its exploration.

The European Space Agency developed the Aurora Programme, which was explicitly presented as a sequence of robotic missions which would lead to human Mars exploration. India has succeeded in sending a robotic orbiter to Mars, and China has started a series of robotic lunar exploration missions. These last two countries did not explicitly state an intention to send human Mars explorers, but possibly aim at acquiring apabilities to eventually reach this goal.

Meanwhile the ISS showed that effective international cooperation is possible in space for complex and large enterprises developed over decades. The idea of a human Mars mission performed by a truly international group, even larger than that engaged in the ISS, has gained acceptance.

***Lesson**: The recent arrival of private enterprises in the field of space activities, whose official purpose is landing humans on Mars as a first step toward colonization, is the begginging of an era where space explorations will no longer depend solely on Governmental organizations. These organizations should wisely support private initiative and make use of them in order to achieve the goals of space exploration in a leaner and more efficient way.*

Chapter 3
INTERNATIONAL COOPERATION

3.1. Introduction & Purpose

This chapter develops a rationale for why the human exploration of Mars should be an international endeavor. An examination of the benefits of international cooperation in space and an overview of recent partnerships lead to observations and recommendations to develop an international effort for the human exploration of Mars. *International cooperation* is used here as the involvement of two or more countries working together to execute a single mission or series of missions. This is distinguished from *international coordination*, where two or more countries keep each other informed of their complementary activities in their respective execution of a broad enterprise [1].

3.2. Why International? A follow-on to the IAA International Exploration of Mars 4th Cosmic Study

The IAA 4th Cosmic Study *International Exploration of Mars* [2], provided a rational for why international cooperations should be a defining feature of any human exploration of Mars in the third chapter of "*Why International?*". The rationale combined philosophical, technological, financial, scientific and educational considerations with a pragmatic observation: "there are few nations ... that might be able to afford such expenditures ... economic and political variables make a unilateral human Mars expedition very unlikely for the foreseeable future." The second section of this chapter provides its own examination of the benefits of cooperation in space exploration.

The IAA 4th Cosmic Study was submitted for release in August 1996, one month after the landing of NASA's Mars *Pathfinder* Lander and of *Sojourner*, the first successful Mars rover. The success of Mars *Pathfinder* gave new momentum to the robotic exploration of Mars and several space agencies have since sent orbiters and surface probes to Mars. Despite the nearly 20 years of Mars robotic exploration that followed, this study recommendation very much echoes that made by the IAA 4th Cosmic Study:

An International Mars Exploration Program, including human missions, is the next step in a series of what has been mostly nationalistic explorations.... An ambitious robotic Mars program is already underway, based on national programs, but with international contributions.... This document urges the extension of this cooperative international robotics program into a formal International Mars Exploration Program which includes human missions.

The 4th Cosmic Study further recommended that informal, non-official consultations and conferences under the auspices of an organization, which the report called International Mars Exploration Forum (IMEF), be established to define the technical and political issues for an international cooperative exploration program. This forum would then evolve in stages to a formal activity, as the International Commission for the Exploration of Mars (ICEM), to set the international agreements needed for achieving human presence on Mars. The third section of this chapter provides an overview of recent international cooperation efforts, including the Global Exploration Strategy (GES) and the ISECG, which are fulfilling the coordination role of the IAA's International Mars Exploration Forum. The final section proposes a series of steps for establishing an international cooperation framework suited to the human exploration of Mars.

3.3. The Necessity of International Cooperation for the Human Exploration of Mars

International cooperation has been promoted as an enabler of space exploration goals based on the premise that "sustainable space exploration is a challenge that no one nation can undertake on its own." [3]. Table 3.1 provides a list of the most-often cited benefits of international cooperation based on a literature search. The table lists all references mentioning benefits. These are mostly documents from American and European organizations and authors, or reports from international organizations such as the IAA.

Notwithstanding the limitations mentioned above, several observations can be made from the results of this search. A wide spectrum of cooperation benefits is reported by all American, European, and international organization references. These range from pragmatic considerations of cost sharing and resource pooling, to more qualitative considerations of image and moral obligation. The pragmatic reasons are most often cited and they directly apply to the human exploration of Mars due to the magnitude of the investments and the breadth of technologies and capabilities needed to achieve this goal. However, as Mr. Jean-Jacques Dordain, at the time Director General of ESA, reflected, "Ultimately cooperation for the sole sake of saving money can prove disappointing" [18]. Some partners of the ISS have expressed a sense that international cooperation is effective as a means to achieve space goals but not as one to save money [13].

	US Perspective	European Perspective	Perspective of intl. organizations
Cost sharing & leveraging of investments	[1], [4], [5], [6], [7], [8], [9]	[10], [11], [12]	[2], [14], [15]
Access to enabling capability	[4]	[11], [13]	
Expanded technological pool	[1], [4], [5], [7]		[2], [15]
Enhanced capability	[1], [4], [7]	[10]	[3], [14]
Expanded user base Exchange of data	[7]	[13]	
Improved program robustness	[6], [8]	[10], [12]	[14], [16]
Acceleration of schedule	[5]		[15]
Strengthened intl. relationships. Promotion of intl. policy objectives	[1], [4], [5], [8]		[2], [14]
Enhanced global security	[1]		[2], [3]
Improved nation's image	[1], [6], [8]	[10]	
Workforce stability	[5]		
Philosophical/ moral motivations		[11], [12]	[2]

Table 3.1. List of the most-often cited benefits of international cooperation based on a literature search.

The literature search returned a large number of US papers. Besides the inevitable bias in the search (performed in the US by a US citizen), this fact may indicate that although NASA has a clear mandate to expand international cooperation [17], there is a continuing debate within US space stakeholders to rationalize international cooperation simply as a means of achieving US space exploration goals. In some instancies, the US still prefers to explore alone. As an example, the Inspiration Mars project is an American non-profit organization whose mission to send humans on Mars fly-bys is tagged "Mission for America". In contrast, cooperation is essential for all ESA partners to achieve their goals in space. In this literature search, few European references were found. This may be due to the fact that in Europe there is no alternative to cooperate in large scale projects, and therefore no need to argue for and rationalize its benefits.

One perspective not represented in Table 3.1 is that of emerging space nations. The IAA Head of Agencies and International Space Exploration Forum held in Washington D.C. on January 9, 2014 brought together representatives from thirty-two countries and gave them an opportunity to voice their views on cooperation. At this event, several representatives from emerging space nations expressed a desire to cooperate in space as a means to develop their local industry but at the condition that cooperation took place on a level playing field, evolving from seller-buyer relationships to genuine technology transfers from established space nations to emerging ones.

In summary, international cooperation is seen by established space nations as an enabler of *sustainable* human space exploration because it improves affordability by each partner, returned value (enhanced capability) to all, safety, and program robustness. International cooperation may not be needed for a one-off human mission to Mars (e.g. Inspiration Mars) or a sustained robotic exploration of Mars, but is seen as necessary for realizing the goal of sustained human Mars exploration, which requires a much larger budget and gamut of technologies. Cooperation is also an opportunity for younger space agencies to participate in exploration provided that there is a mutual benefit among all partners.

3.4. Status of International Cooperation in Human Space Exploration

3.4.1. Examples of Space Projects with Major Cooperation

International Space Station (ISS)

The ISS program is currently the best model of international cooperation in space. Indeed, the strong and tested working relationship among its partners is considered perhaps the most important outcome of the ISS program [4]. The importance and value of the ISS partnership was made clear recently by the fact that it remained one of the very few on-going working relationships between the US and Russia during the geopolitical tensions involving the two countries in early 2014. There is a wide consensus that the ISS management framework is the right starting point to expand international cooperation models to mission campaigns beyond LEO.

As the ISS was nearing completion, the Multilateral Coordination Board released a summary report to document lessons learned from the ISS experience up to that point and to indicate how they may apply to future human exploration programs operating beyond LEO [19]. The document contains more than fifty lessons learned covering many areas including mission objectives, space architecture, international partner structure and coordination, and commercial involvement. Each ISS-specific lesson learned is accompanied by a guideline on how it may apply to exploration beyond LEO. In building its GER (see 3.4.2), the ISECG intended to apply as many ISS lessons learned as possible but found a subset of important lessons that could not be readily implemented on exploration missions (e.g. realistic expectations, appropriate interdependence, and redundant transportation commitments) [20]. The Workshop on International Cooperation for Sustainable Space Exploration also found the direct application of the ISS cooperation model to exploration to be questionable [21]. Future initiatives for cooperative space exploration will need to analyze these lessons learned further and to minimize the residual risks associated with them not being applicable to exploration in a practical way.

In summary, the international cooperation model established for the development and utilization of the ISS, while very successful, is not readily applicable to the case of an international human exploration of Mars. A dedicated effort needs to take place to create a partnership model that addresses the partnership risks intrinsic to space exploration.

ExoMars and Crew Space Transportation System

While international cooperation has the potential to provide all the benefits mentioned in previous sections, it must at the same time be seen as a risk that if not managed effectively can be detrimental, cause schedule delays, or worse, lead to a premature end of the program. Recent cooperation efforts between Europe and the US illustrate this point. In February 2011, the US pulled out from a planned joint mission to Jupiter and in February 2012 the US pulled out of the ExoMars mission two and half years after approving the Joint Mars Exploration Initiative [22]. In the case of ExoMars, the termination of the partnership impacted significantly on technology development, funding, and schedule which transformed the ExoMars program into an unexpected ESA-Roscosmos cooperation.

In the area of human spaceflight, an ESA-Roscosmos joint effort on the Crew Space Transportation System (CSTS) was prematurely ended after the initial concept study made apparent

- The difficulties of setting appropriate management structures between the two agencies and their respective industrial partners and
- The challenges of establishing a working collaborative environment given the hurdles of export control regulations [23].

In summary, international agreements to cooperate on space projects are rarely legally binding and cooperation relies on a sense of solidarity and common purpose among partners. These are sometimes overshadowed by financial priorities and restrictive regulations.

3.4.2. The Coordination of Global Exploration

Global Exploration Strategy

In 2006, fourteen space agencies began developing a vision for peaceful robotic and human space exploration and a list of common key space exploration goals. This vision was articulated as The GES, which serves as a non-binding framework for international coordination to support a future space exploration beyond LEO [3].

The ISECG was established in response to the GES call for a coordination mechanism by which nations can share plans for space exploration and collaborate to strengthen both their own projects and collective ones. The purpose of the ISECG is to develop the Global Exploration Strategy by:

- Providing a forum for participants to discuss their interests, objectives and plans in space exploration.
- Promoting interest and engagement in space exploration activities throughout society worldwide.
- Developing non-binding findings, and consensus-based recommendations [16].

Most recently, the International Objectives Working Group of the ISECG produced a paper, entitled “Benefits Stemming from Space Exploration”, which provides the view of the ISECG on the concrete and qualitative benefits of investing in space exploration. This ISECG paper will help space agencies from emerging nations in promoting the benefits of exploration and rationalize the return on investments made in space to their constituencies. This will build public support for participating in future space exploration cooperation opportunities.

In addition, the ISECG Exploration Roadmap Working Group and International Architecture Working Group produced an update to the GER [15]. The updated roadmap reflects the vision of the ISECG for an evolutionary exploration path from the existing ISS to the common goal of sustainable human missions to Mars. The roadmap aims to be a single reference mission scenario; however, at this early stage of the coordination process, it has the feel of an ad-hoc collage of the partners’ individual plans, with some aiming for Near-Earth Objects (NEO) and others for the Moon as intermediate steps. The GER is still an extremely valuable product of the ISECG as it forms the technical basis for informing the necessary future cooperation agreements between agencies and governments.

It is not yet clear from the roadmap how the ISS cooperation model will be evolved and expanded to support human exploration of Mars. The fact that agencies are considering parallel exploration paths (NEO or Moon) in the short term suggests that the space exploration model of international cooperation is at risk of not being drafted until a second phase of exploration. The space exploration cooperation model needs to have its own "evolutionary strategy" like the one the ISECG developed for maturing technologies needed for Mars exploration.

3.5. Role of Commercial Space: International Industry Team

Since 2009, international space industry partners from Japan, Canada, Russia, Europe and the US have been working together to study the next steps in space beyond the International Space Station. This group is the industry counterpart of ISECG in that it aims at developing global exploration scenarios and at identifying contributions consistent with each other's areas of expertise, technology development priorities, and other national interests [24].

As mentioned earlier in the case of the CSTS program, a successful industry-to-industry dialog is key to the success of a cooperative effort at the agency-level. Initiatives like the International Industry Team are valuable because they begin the process of establishing working relationships between employees of international space companies who will continue to work together in the future. Plans to develop a cooperation model for space exploration must engage the private sector and leverage their expertise and lessons learned.

At least in the case of the United States, but also partially in the case of Europe, the new trend is using private companies to grant access to orbit also to space agencies. Along these lines Space X is developing a heavy lift launcher specifically intended for human Mars missions, and other companies are working to reduce the cost of orbiting payloads, a critical factor in designing affordable human Mars missions.

Together with what can be defined 'semi-private' access to space (launch services managed by private companies for space agencies who 'pay the ticket' for their astronauts) a completely private approach to space exploration has recently been proposed [25] in which space exploitation, but also exploration, is entirely performed by private entities rather than by Space Agencies.

While for decades, nongovernmental organizations like the Planetary Society or the Mars Society advocated space exploration and human Mars exploration in particular. In recent years other non-profit foundations, like Mars One or Inspiration Mars, declared their intention of launching Mars Exploration missions. There is little doubt that in the future other private foundations or companies will follow suit; however, these will not be dealt with in this study, which focuses on human Mars exploration as an international endeavor, mostly performed by space agencies. That said, these private initiatives may well impact the subject of the study, since:

- Private managing of orbital services promises to cut the cost of reaching LEO. This will make it easier to mount a human Mars exploration, with two possible opposite outcomes, namely:
  - Making a larger Initial Mass in Low Earth Orbit (IMLEO) affordable, allowing lower performance but proven chemical propulsion technology.
  - Allowing cheaper development of advanced nuclear (thermal or electric) propulsion.
- Private missions would increase the interest of the public, enhancing the chances of success of proposals to explore Mars. Significantly, the foundation proposing the Mars flyby mission has the name 'Mars *Inspiration* Foundation'.
- With a growing economy and an expanding private space sector, it is likely that, faced with governmental indecision about human Mars missions, private entrepreneurs will take it upon themselves to complete the task space agencies fail to fulfill.
- Exploitation of space resources (perhaps asteroids) by industry may eventually allow the private space sector to grow sufficiently to mount a Mars exploration program funded privately or by a joint private-public enterprise. For instance, mining Mars satellites is conceivable, with industry offering space agencies transportation service all the way to Mars.
- Any optimism about private enterprise entering the space arena must be tempered by the fact that companies work for profit, and there must be a 'Mars market' separate from Governmental projects justifying initiatives or participation in Mars missions. So far, only a modest LEO market has materialized (e.g., "Blomberg News", March 10, 2015).

3.6. Path to an Effective International Mars Exploration Framework

The previous sections of this chapter are summarized in three observations:

- Effective international cooperation among space faring nations is necessary to achieve the shared goal of a *sustainable* human exploration of Mars;
- The international cooperation model established for the development and utilization of the ISS, while very successful, is not readily applicable to the case of international human exploration of Mars;
- International cooperation introduces risk. Ineffective international cooperation efforts have caused major disruptions to, and in some cases the premature end of, recent space programs.

To conclude, a successful program for sustainable human exploration of Mars depends on developing and testing an international cooperation framework adapted to space exploration. In this view, the effort needed to set up a working international cooperation framework is no different than that needed to mature a new technology critical to successful execution of a mission. The development and maturation of such international cooperation is an integral piece of any GER

The high level recommendation of this study group to partners of GES and ISECG is to start collectively defining a cooperation framework that supports and informs the execution of the GER. More specific recommendations are made in the following subsections.

3.7. Motivation for a global international cooperation,

The first recommendation is for the ISECG members to document and introduce the reasons why they consider international coordination and cooperation an effective means to achieve their space exploration goals in future revisions of the GER. This will have two benefits. First, it will help partners understand each other's motivations for joining the ISECG and appreciate each other's perspective on international cooperation, an important lesson learned from past cooperation programs [22]. Second, it may motivate other nations to join the ISECG.

A human Mars mission covers so many aspects of science and technology that some areas can be covered by partners without a long standing history in space exploration. Thus, countries which cannot yet be defined as 'spacefaring' or which are just now starting their journey into space can get involved.

Organizations having a true international standing, like the IAA, the Committee on Space Research (COSPAR), the International Astronautic Federation (IAF) and others may be instrumental in involving the emerging spacefaring or emerging nations in this enterprise.

The Human Space Technology Initiative (HSTI) launched in 2010 by the Office for Outer Space Affairs under the framework of the United Nations Programme on Space Applications, is an example of an initiative aimed at involving more countries in activities related to human space flight and space exploration. It stresses increasing the benefit from the outcome of such activities through international cooperation to make space exploration a truly international effort.

The role of the Initiative in these efforts consists of providing a platform to exchange information, to foster collaboration between partners from spacefaring and non-spacefaring countries and to encourage emerging and developing countries to take part in space research and to benefit from space applications.

The Initiative is based upon three pillars:

- **International Cooperation:** To promote international cooperation in human spaceflight and activities related to space exploration;

- **Outreach:** To promote increased awareness among Member States of the benefits of utilizing human space technology and its applications;
- **Capacity-building:** To build capacity in microgravity science education and research.

3.8. Definition of the International Mars Exploration Framework

The second recommendation is to form a group tasked with developing criteria for creating a formal framework for international cooperation in space exploration (ISS Lessons Learned #35, [19]). The responsibilities of this group match that of the International Commission for the Exploration of Mars envisioned by the IAA 4th Cosmic Study. It could be a new working group of the ISECG responsible for defining the intergovernmental agreements and the operational plans by which the Mars exploration program will proceed. The Mars exploration framework has several issues to define [2]:

- Intergovernmental agreements
- Charter and organizational structure
- Funding plans and protocols
- Official project language
- Technical and legal standards
- Choice of technologies
- Technology transfer and spin-off policy
- Participation of emerging space nations
- Role of commercial space
- Intellectual property rights

The next section expands on the concept of framework dimensions and how they are used to measure how a given program of missions supports the development and maturation of the cooperation framework.

3.9. Roadmap for Implementing the Framework

The third recommendation is to task the above group with developing the roadmap for implementing the international cooperation framework. This roadmap will identify the steps needed to mature each dimension of the framework. An example of such a roadmap for human exploration capability development is in Figure 3.1 [26].

It is possible to introduce the concept of Human Mars Mission Feasibility Index (HMMFI). It can include technical, human, programmatic, political, and sustainability parameters which define the the feasibility of a human Mars mission. If the value of this index is updated every year, the evolution of the different parameters can be monitored and the progress toward a human Mars mission can be assessed. By comparing the value of the HMMFI computed for different mission architectures it is possible to assess which one is the most likely to be implemented.

The concept of HMMFI is further discussed in Appendix D.

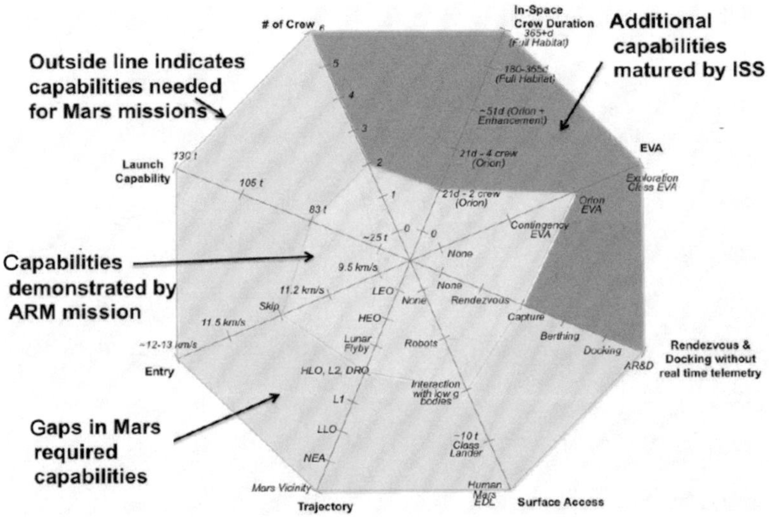

Figure 3.1. Example of roadmap for human exploration capability development
The figure identifies all the steps needed for each capability (dimension) to evolve from its current state of the art (human spaceflight in LEO) to the maturity level needed by human exploration of Mars.

3.10. Synergy between the Evolution of the Cooperation Framework and the Global Exploration Roadmap

The final recommendation is for the above group and other working groups of the ISECG to assess how GER missions contribute to the advancement of the cooperation framework, not only how they can contribute to the development of technologies needed for the human exploration of Mars. This assessment can shape the portfolio of missions selected by the roadmap and increase the synergy among the different groups. For example, on top of scientific and technological benefits, an international Mars Sample Return mission could prove valuable as a testbed for a space exploration cooperation framework among partners interested in human Mars exploration.

Also, the concept of stepping stones must be further developed if the current goal is Mars. Intermediate goals must be evaluated based on their consistency with the stated Mars objective so that each new mission becomes a stepping stone with known technical and programmatic contributions towards the long term objective of the human Mars exploration.

Chapter 4
THE ENVIRONMENT

These are the four environments relevant to a Mars mission [1]:

- Low Earth orbit
- Interplanetary space
- Mars surface
- Mars satellites

Since the present study deals mainly with a mission to the The surface of Mars, the problems related with forward and backward contamination are of paramount importance and are linked with the environment astronauts will find on the planet. A very important point is whether the Mars environment has hosted life in the past or still supports living creatures. This point is essential in the decisions to be taken about contamination of both types, and will strongly influence the activities humans will perform once the planet has been reached.

4.1. Low Earth orbit environment

Low Earth Orbits (LEO) are those with heights between 160 and 2000 km, so that the related environment is relevant only for the early stages of a Mars Mission; in particular if the spacecraft is assembled in orbit

The Earth's magnetosphere protects our planet and all space below the Van Allen belts from most solar and galactic radiation. Owing to this, radiation is moderate in LEO, even during strong solar activity (solar flares, Coronal Mass Ejections (CMEs)). However there is an anomaly in the Earth's magnetic field off the coast of Brazil and in that zone much stronger radiation reaches the upper atmosphere. The Van Allen belts are symmetric about the Earth's magnetic axis, which is tilted with respect to the Earth's rotational axis by an angle approximately eleven degrees. This tilt, together with an offset of about 450 km causes the inner Van Allen belt to be located closer to the Earth's surface over the South Atlantic Ocean and farther from the Earth's surface over the North Pacific Ocean, thereby producing an enhanced, localized radiation feature called the South Atlantic Anomaly (SAA).

The boundaries of the SAA vary with altitude and its shape changes over time. At an altitude of 500 km, the SAA ranges between -90° and +40° in longitude and -50° to 0 in latitude. Its extent increases with increasing altitude.

The characteristics of the upper atmosphere and of the space above it are quite variable, both with respect to altitude and time. Between 90 and 1000 km the average values of pressure, density and temperature reported by the US Standard Atmosphere (an extension of the International Standard Atmosphere) can be used [2], [3].

Atoms in the upper atmosphere may be multiply ionized and the atmospheric layer between 50 and 600 km altitude is referred to as the ionosphere. Ions are mostly oxygen ions, but above 300 km the composition changes gradually to mostly hydrogen ions.

The conditions in that zone of space are well known, but are very variable depending on "space weather", a term commonly used to define the phenomena involving ambient plasma, magnetic fields, radiation and matter in a region of space. The changing space weather close to the Earth is a consequence of the behavior of the Sun, the interaction of solar emissions with the Earth's magnetic field, and our location in the solar system.

As a whole, the solar system space weather is greatly influenced by the speed and density of the solar wind. Data on the current solar wind speed and density and on solar flares and CMEs are continuously monitored.

The solar wind density peaks with the solar activity cycle approximately every eleven years. During these enhancements, satellite drag increases and the danger to a spacecraft of losing altitude and deorbiting increases. To remain in LEO between 200 and 400 km satellites require periodic reboosting, particularly if their surface/mass ratio is large. The ISS is subject to high atmospheric drag owing to its large solar panels. Reboosting becomes even more important during periods of high solar activity.

At higher altitudes, pressure and density decrease quickly; at about 1000 km a satellite may remain in orbit indefinitely, at least with reference to the normal service time of man-made machines.

Apart from plasma, space is full of debris of various types, both natural and artificial. Natural debris consists mainly of very small meteoroids, micrometeoroids and dust grains that enter the Earth's atmosphere where they are destroyed by air drag. Occasionally, larger meteoroids reach the terrestrial surface.

Most space debris in LEO are, however, artificial. Large pieces of such debris are accurately tracked by radar and telescopes so that their orbits are well known. Although new debris is always being produced, the older items decay owing to atmospheric drag and eventually re-enter the atmosphere, where they are completely destroyed before reaching the ground.

The eleven year solar cycle has a strong effect on space debris since the density of the high atmosphere is much greater near to solar maximum conditions than at solar minimum. A periodic clean-up of debris in the lowest orbits thus occurs.

Smaller debris are produced by the accidental or intentional explosion of upper propulsion stages or satellites. About half of the centimeter-sized debris has been estimated to be produced in this way. International treaties forbidding intentional explosion of satellites are being prepared; they state that precautions must be taken against accidental events which may produce space debris.

The most critical orbits are those orbiting between 1000 and 1400 km, where drag is insufficient to cause debris to re-enter. In the case of the ISS, a dangerous piece of debris identified in advance can be avoided by maneuvering. Very small objects do not cause damage and debris smaller than a few millimeters are stopped by the ISS shielding. The most dangerous debris are those which cannot be detected from the ground and are larger than a few millimeters. The probability that a one centimeter sized particle will pierce the ISS hull is about 1% over its 20 years life. The danger of such an accident occurring during the assembly of a spacecraft for Mars is small but must be taken into account.

There are well consolidated design practices for commercial and scientific satellites in LEO, so this can be considered to be an environment that does not pose unexpected problems.

4.2. Interplanetary space

Space beyond the Van Allen Belts is crossed by the solar wind which fills the whole solar system [1]. The solar wind is mainly composed of hydrogen ions (protons) flowing at high speed out of the Sun's corona. The temperature of the corona is so high that the coronal gases are accelerated to a velocity of about 400 km/s. This component of the solar wind, the so called 'slow solar wind', has a temperature of 1.4 - 1.6 $\times 10^6$ K and a composition similar to that of the corona. Over coronal 'holes' the speed of the solar wind can reach 750 - 800 km/s and show a temperature of about 8×10^5 K. The composition of this 'fast solar wind' is close to that of the Sun's photosphere. From the colder outer layers, the solar wind can be as slow as 300 km/s [4].

While the slow solar wind is mostly ejected from the Sun equatorial region (up to 30° - 35° latitude), the fast solar wind originates from coronal holes located mostly near the Sun's magnetic poles.

The interaction of particles of different velocities with the Sun's rotation causes the solar wind to be quite unsteady and the space weather in the whole solar system is very variable. From time to time, fast-moving bursts of plasma called Interplanetary Coronal Mass Ejections (ICMEs) may disrupt the standard pattern of the solar wind through propagating in the surrounding space electromagnetic waves and fast particles (mostly protons and electrons) to form showers of ionizing radiation. When these ejections impact the magnetosphere of a planet they temporarily deform that the planet's magnetic field. On Earth they induce large electrical ground currents and send protons and electrons toward the polar atmosphere, where they form aurorae.

Solar flares and coronal mass ejections constitute a danger to spacecraft, with or without humans on board. Since they are still unpredictable, crewed interplanetary spacecraft must incorporate a radiation shelter. The crew must enter the shelter in case of dangerous solar events and stay there during the hazardous time.

Owing to the motion of charged particles, an Interplanetary Magnetic Field (IMF) pervades the whole solar system.

The interplanetary medium is also filled with galactic cosmic radiation from our galaxy. Cosmic radiation impinging on the Earth's atmosphere consists mostly of protons (90%), plus about 9% helium nuclei (α particles) and about 1% electrons (β particles). The particle energies are in the range above 1000 eV.

Galactic Cosmic Rays (GCRs) from outside the solar system are atomic nuclei trapped by the galactic magnetic field. These have their surrounding electrons stripped away during their travel through the galaxy at close to the speed of light. Their energy spectrum may reach to TeV or PeV, although at very low fluxes. As they travel through the very thin gas in interstellar space they emit gamma rays. Their atomic composition is similar to that of the Earth and solar system.

Another component of the cosmic radiation are Anomalous Cosmic Rays (ACRs). These are neutral atoms of interstellar matter flowing through the solar system at about 25 km/s (charged particles are kept outside the heliosphere by the interplanetary magnetic field). When closer to the Sun, these atoms undergo the loss of one electron by photo-ionization or by charge exchange, and are then accelerated by the Sun's magnetic field and the solar wind. ACRs include helium, oxygen, neon, and other elements with high ionization potentials.

Apart from these heavy particles, there is also the cosmic microwave background radiation, consisting of very low energy photons (about 2.73 Kelvin) which are remnants from the time when the universe was only about 380,000 years old. Neutrinos, photons of different energies (produced by the Sun, other stars, quasars, black-hole accretion disks, gamma-ray bursts and so on), electrons, and other particles are also present.

All these particles are not dangerous on Earth as they are deflected by the Earth's magnetic field or are stopped by the atmosphere. In interplanetary space the radiation dose they inflict on a human crew depends on the duration of exposure; it has been estimated that a Mars mission lasting one year would result in a dose of about 1 Sievert (Sv). For comparison, the background average dose is 6 mSv/year. A 1-Sv dose entails a significant cancer risk. Apart from plasma, there a tiny amount of neutral hydrogen is also present: at the distance of the Earth's orbit from the Sun, the concentration of neutral hydrogen is about 10^4 atoms per m^3. Some of these atoms come from interstellar space.

A relatively small amount of dust particles - micrometeoroids - exist in the solar system. Much of this dust is thought to have been produced in collisions between asteroids and spread through the shedding of material by comets while passing close to the Sun. About 30,000 tons of interplanetary dust particles are estimated to enter the Earth's upper atmosphere annually.

The vacuum is much higher than in LEO and hydrogen ions from the Sun substitute for oxygen ions from the Earth's atmosphere. So, while the environment in LEO is oxidizing, that in deep space is reducing.

Since Mars has no magnetosphere, the environment in Mars orbit is similar to that in interplanetary space.

4.3. Mars surface

The main features of Mars are reported in Table 4.1. Mars has the largest volcano in the solar system, Olympus Mons, 25 km high (but there is no active volcano on Mars now), and a canyon, Valles Marineris, which is likely the deepest and widest in the solar system. Mars bears the traces of impressive and dramatic events in the past, which have remodeled its surface – including the northern lowlands, Vastitas Borealis, which was probably formed due to the impact of a large meteorite, the Tharsis Bulge, which is probably of volcanic origin, three huge volcanoes (Pavonis, Arsia and Ascraeus Mons), and a huge number of impact craters, chasms, and mountains. If Vastitas Borealis is considered to be an impact basin, it is the largest found in the solar system, four times the size of the lunar South Pole-Aitken basin.

The poles are covered by ice caps which shrink in summer and grow in winter. The northern ice cap is mainly water ice while the southern cap has a frozen carbon dioxide upper layer and an underlying layer of water ice. About 25% to 30% of the planet's atmosphere condenses during a polar winter, forming thick slabs of CO_2 ice, sublimating again when local the pole is exposed to sunlight. This creates huge wind storms from the poles with wind velocities up to 400 km/h but low dynamic pressure. It is estimated that up to two million cubic kilometers of water ice may be contained in the northern ice cap.

	Mars	Earth
Mass (10^{24} kg)	0.64185	5.9736
Volume (10^{10} km³)	16.318	108.321
Equatorial radius (km)	3396.2	6378.1
Polar radius (km)	3376.2	6356.8
Ellipticity (flattening)	0.00648	0.00335
Topographic range (km)	30	20
Mean density (kg/m³)	3933	5515
Surface gravity (m/s²)	3.71	9.81
Surface acceleration (m/s²)	3.69	9.78
Escape velocity (km/s)	5.03	11.19
Solar irradiance (W/m²)	589.2	1367.6
Orbit semimajor axis (10^6 km)	227.92	149.60
Sidereal orbital period (days)	686.980	365.256
Perihelion (10^6 km)	206.62	147.09
Aphelion (10^6 km)	249.23	152.10
Synodic period (days)	779.94	
Mean orbital velocity (km/s)	24.13	29.78
Max. orbital velocity (km/s)	26.50	30.29
Min. orbital velocity (km/s)	21.97	29.29
Orbit inclination (deg)	1.850	0.000
Orbit eccentricity	0.0935	0.0167
Sidereal rotation period (hrs)	24.6229	23.9345
Length of day (hrs)	24.6597	24.0000
Obliquity (deg)	25.19	23.45
Min. dist. from Earth (10^6 km)	55.7	
Min. dist. from Earth (light minutes)	3	
Max. dist. from Earth (10^6 km)	401.3	
Max. dist. from Earth (light minutes)	22	

Table 4.1. Main features of Mars compared to those of Earth along with the surface acceleration at the equator, taking into account the planet's rotation

Differences between the two ice caps are due to the orbital eccentricity and orbital axis inclination of Mars, both of which are higher than is the case on Earth. This combination makes the seasons far more extreme in the southern than in the northern

Martian hemisphere. Mars nears perihelion when it is summer in the southern hemisphere and nears aphelion when it is winter. This is believed to be the explanation for the occurrence of violent dust storms which last for months at a time.

The Martian day, usually referred to as a 'sol', is slightly longer than the Earth's day (24 h, 39 min, 35 s). The tilt of the rotation axis is similar to that of Earth (25° and 23°, respectively), producing seasons similar to ours, although they last longer, owing to the much longer Martian year (687 Earth days).

Mars is smaller than Earth, with surface area about as large as the sum of all the continent areas of our planet. A simplified map of Mars is presented in Figure 4.1.

Figure 4.1. Simplified map of the Martian surface (from a NASA image).

The Martian atmosphere is very thin: the pressure at the ground is less than 1% of the atmospheric pressure on Earth (it roughly equals atmospheric pressure on Earth at 35 km altitude) and varies significantly with altitude and latitude, from a minimum of around 0.3 millibar on Olympus Mons to over 11.6 millibar in the depths of Hellas Planitia, with a mean surface level pressure of 6.36 millibar, changing with the seasons from 4.0 to 8.7 millibar. The average density on the ground is about 0.020 kg/m^3. This variability poses challenges in planning spacecraft re-entry.

The composition by volume of the atmosphere is reported in Table 4.2

gas	percentage
carbon dioxide	95.97%
argon	1.93%
nitrogen	1.89%
oxygen	0.15%
carbon monoxide	0.06%
water	210 ppm
nitrogen oxide	100 ppm
neon	2.5 ppm
heavy water	0.85 ppm
krypton	0.3 ppm
methane	traces

Table 4.2. Average composition of Mars atmosphere

Methane is concentrated in a few places during the northern summer. Since methane is broken down by ultraviolet radiation, in order for it to be present a mechanism to produce it is required. Explanations include volcanic activity, cometary impacts and the presence of methanogenic microbial life forms. The mean molecular weight of the atmosphere is 43.34 g/mole. Clouds of water ice were photographed by several probes,

The fine dust in the atmosphere give the Martian sky a tawny color when seen from the surface.

At the Viking landing sites, wind speeds of 2 to 7 m/s were recorded in the summer, 5 to 10 m/s in the fall, and occasionaly 17 to 30 m/s during dust storms.

Although the wind speeds are high, the aerodynamic forces exerted by the wind are low owing to the low atmospheric density: vehicles and structures on the Martian surface are not expected to be mechanically stressed by the strong winds.

Winds carry large quantities of dust rich in iron oxide which is finer than on the Moon (particulate diameter is about 1.5 μm). This poses dangers to machinery and human beings. Mars plains are frequently crossed by dust devils. These swept clean the solar panels of the Spirit and Opportunity rovers more than once, thereby contributing to maintain them operational for longer than was expected.

The average temperature on the ground is −63 °C, with marked variations with the time of the day and of the year. At the site of the Viking 1 landing diurnal variations between −89 °C and −31 °C were recorded. Larger variations, from −120 °C to −14°C were recorded during the years-long Viking missions. During the southern hemisphere summer, temperatures as high as 20 to 30 °C above zero have been recorded. Liquid water cannot exist on the surface at these temperatures and pressures and most of the frost depicted in the images from the Viking landers is frozen carbon dioxide.

The Martian atmosphere offers very little protection from the Sun's ultraviolet radiation and there is also limited protection from cosmic rays due to the almost complete absence of a planetary magnetic field after Mars lost its magnetosphere about four billion years ago. Heavy GCR particles reach down to the surface and striking the iron oxide in Martian rocks release high energy alpha particles, which are not stopped by the thin atmosphere. From the point of view of radiation risk, Mars presents only a slightly better place to be than on the Moon, even though the atmosphere scatters light almost like that on Earth.

There is evidence that at the beginning Mars had plate tectonics and a planetary dynamic effect, producing a global magnetic field. Some remnants are still found in the form of local magnetization.

The geography of Mars is complex. The main features are shield volcanoes, lava plains located mostly in the northern hemisphere, and highlands with a large number of impact craters and deep canyons. The four largest volcanoes are all extinct. A total number of 43,000 craters with a diameter of 5 km or greater have been found, together with a large number of smaller ones.

The largest canyon, Valles Marineris, has a length of 4000 km and a depth of up to 7 km. It was formed due to the swelling of the Tharsis area which caused the crust in the area of Valles Marineris to collapse. Also Ma'adim Vallis is a canyon much bigger than the Grand Canyon of Colorado.

Mars Odissey and the Mars Reconnaissance Orbiter sent pictures of entrances to large caves, 100 m to 250 m wide; moreover it is possible that lava tubes exist that are larger than those on Earth because of the lower gravity. Caverns and lava tubes may provide shielding from the micrometeoroids, UV radiation, solar flares, and high energy particles that bombard the planet's surface. They are also are good targets when searching for liquid water, signs of life, and possible locations for human settlements.

Martian rocks seem to be mostly basaltic, although a portion of the Martian surface seems to be richer in silica than is typical of basalt. The plains are rugged, similar to rocky deserts on Earth, covered by red sand, with rocks and boulders scattered all around. In general, it is difficult to land close to the most interesting places; among these are the steep slopes of mountains and canyons, which are quite difficult to negotiate by wheeled or tracked vehicles.

The soil is essentially regolith, rich in finely-grained iron oxide dust. Its granularity and composition is highly variable from place to place, due to wind and water erosion in ancient times when water was flowing on the surface.

The results of the experiments conducted by the Viking probes, although highly controversial, seem to exclude the possibility of finding life-forms on the surface of the planet. Even if some life-form might be found in the future, in particular in places shielded from radiation and direct sunlight, like the bottom of canyons or caves, it is safe to say that biological products are not a constituent of the surface of the planet.

This may make surface mobility easier, since on Earth the most difficult terrains are those rich in products of biological origin, where vehicular tracks instead of wheels are often needed.

The geological history of Mars is subdivided into three main periods, namely

- Noachian epoch (named after Noachis Terra), from 3.8 billion to 3.5 billion years ago.
- Hesperian epoch (named after Hesperia Planum): 3.5 billion years ago to 1.8 billion years ago.
- Amazonian epoch (named after Amazonis Planitia): 1.8 billion years ago to present.

The Tharsis bulge formation and extensive flooding by liquid water are ascribed to the Noachian epoch. Extensive lava plains are deemed to have been formed in the Hesperian epoch, while Olympus Mons formed during the Amazonian epoch, along with lava flows elsewhere on Mars.

Owing to the in situ observations performed by robotic spacecraft, it is now certain that in ancient geological times Mars had extensive water coverage, with liquid water running on the surface in addition to geyser-like water flows. At that time the atmosphere was also much denser.

Large quantities of water are thought to be trapped underground. In the northern hemisphere an ice permafrost mantle stretches down from the pole to latitudes of about 60° and large quantities of water ice have been observed both at the poles and at mid-latitudes. A large release of liquid water is thought to have occurred when Valles Marineris formed early in the history of Mars, thereby gouging out massive outflow channels. A much more recent (5 million years ago) outflow of water is supposed to have occurred when the Cerberus Fossae chasm formed. There are also hints of more recent flows of water on the surface, at least for short periods of time, but these findings are still debated.

The two-way communication delay with Mars is very variable and depends on the relative positions of the two planets (it can fall between 6 and 44 minutes). However, accounting for the relative motions of the two planets and the lack of telecommunication satellites in Mars orbit, the communication window may be further restricted to a few minutes per day. There are also periods when Mars is in conjunction with the Sun, in which case no communication is possible.

4.4. Possibility of the presence of life

The planet was once much more suitable for supporting living organisms than it is today, which does not imply that life formally existed there. If life did exist on Mars in the past, fossils may still exist. The identification of alleged fossil microorganisms in the ALH84001 meteorite, which was blasted from Mars into space by a meteorite strike about 15 million years ago and then landed on Earth, is still controversial.

Searching for life on Mars is one of the most important, and most difficult, goals of Mars exploration.

After the results of the experiments carried by the Viking probes, it was considered extremely unlikely that present life can be found on Mars, at least on the surface, and even less in places where landing is easy. Deep canyons like Valles Marineris, where the scarce atmospheric moisture could gather, would be better places to search; in general the most suitable zones for life can be reached only by ground vehicles, piloted or automatic, with the capability to climb steep slopes [5], [6].

Some scientists think that if life really started on Mars, it could have survived in some particularly suitable places, for instance in the permafrost considered likely to be present in the subsoil of much of the planet. If life still exists, a possibility which cannot be completely ruled out, it must occupy a very marginal component of the planet, or at least of its surface, with large zones in which no trace of it can be found.

The 'follow the water' strategy is at present considered to be the best option, but it is likely that the search will prove to be very difficult; it is possible that, even if fossil life is found, this will happen only after many missions have explored difficult areas of the Martian surface. This possibility must be clearly stated and communicated to the public, to avoid disappointment following a negative result, with severe effects on the continuation of the exploration program. It must be stated explicitly that the main goal of human Mars exploration is not finding Martian life and that exploration can be considered successful even if life is not found.

4.5. Mars satellites environment

Mars has two small, irregularly shaped, moons, Phobos and Deimos, which orbit close to the planet. Their orbital data are provided in Table 4.3. They might be captured asteroids, similar to 5261 Eureka (a Martian Trojan asteroid), but their capture by an almost airless world is difficult to explain. Phobo's orbit is lower than synchronous: it rises in the west, sets in the east, and rises again in just 11 hours. The orbit of Deimos is just above synchronous and thus its apparent speed in the sky is low. Even if the period of its orbit is 30 hours, 2.7 days pass between its rising in the east and setting in the west.

The orbit of Phobos is unstable: it decays owing to tidal forces and it will either crash onto the planet or fragment, producing a ring, in about 50 million years.

	Phobos	Deimos
Orbit's semimajor axis (km)	9378	23459
Sidereal orbit period (days)	0.31891	1.26244
Orbital inclination (deg)	1.08	1.79
Orbital eccentricity	0.0151	0.0005
Major axis radius (km)	13.4	7.5
Median axis radius (km)	11.2	6.1
Minor axis radius (km)	9.2	5.2
Mass (10^{15} kg)	10.6	2.4
Mean density (kg/m^3)	1900	1750

Table 4.3. Phobos and Deimos orbital data.

Phobos and Deimos are commonly believed to be carbonaceous chondrites. Like other asteroids of the same type, they may become a target for mining operations; this may lead to synergies between human Mars exploration and industrial exploitation of its satellites by space agencies or by private organizations.

In the past, several proposals for human exploration of the satellites of Mars, where astronauts could teleoperate robots on the planet's surface, have been put forward, sometimes as an alternative to ground exploration of Mars and at other times as a complementary goal.

Exploration of Mars satellites will not be dealt with any further in the present study.

4.6. Choice of landing site

The choice of landing site does not significantly influence mission design and can be left to a later design stage. In the context of a multiple mission program, the point worth discussing is whether or not the landing site should be the same for all missions (see NASA DRM 5, [7], [8])

The NASA study concluded that a single landing site is a better choice if the ultimate goal of Mars exploration is to prepare for colonization. If this is the purpose, each mission will contribute to building a single well equipped and redundant outpost which will last for a long time.

Choosing a different landing site for each mission is an optimal strategy if the main goal is purely scientific, enabling deeper understanding of the geology of the planet and maximizing the chances to make discoveries, including finding life. If this is the goal, among possible choices are Centauri Montes, Nili Fossae, and Arsia Mons which feature respectively relics from the Noachian, Hesperian and Amazonian epochs.

A compromise between these two alternatives is to land on three sites located at moderate distances from each other and within the range of a pressurized rover left by the first mission. This choice presumes a relatively flat ground, since follow-on mission crews would need to move freely among the sites while carrying new equipment in order to enlarge their exploration range. The downside of this strategy is the relatively small size of each outpost.

4.7. Forward contamination

The term 'forward contamination' is used to designate the possible contamination, (mainly biological), of a planetary environment by human and robotic exploration missions. This problem has been debated since the time of the earliest Mars robotic probes and guidelines for planetary protection are already included in the Outer Space Treaty and also issued by COSPAR. Forward contamination is prevented primarily by sterilizing the spacecraft but, while this is already difficult to achieve in the case of robotic missions, it is even more so in the case of human missions.

In the simplest circumstances, contamination may be due only to dormant micro-organisms that have not yet encountered conditions that permit them to metabolize and reproduce. In this case, contamination due to exploration may be biologically reversible. Also, if needed, it might be possible in the future to remove all the life we have brought to Mars and return the planet to its previous biologically pristine state [9].

The difficulty with contamination is that there is a common perception that, if Mars does not contain indigenous life, contamination is not an issue, and this view could even be actively pursued to the point that the planet may be, in the distant future, terraformed[4]. Contamination would be of relevance only if there is such a thing as Martian life. If such life is discovered, strict anti-contamination practices – even with

---

[44] The term terraforming (coming from science fiction, but now commonly used) refers to operations aimed at modifying the environment of a planet to make it suitable to support terrestrial life forms.

the difficulties they would cause for human exploration – should be applied but, if life is not discovered, then the uncertainty will remain.

A suggestion to fit the case where no life is found is to divide the surface of the planet according to two possible situations namely:

- That in which we are sure there is no life, where less strict precautions can be taken to make exploration easier, and
- That in which life might be present, where stricter rules should apply.

In the latter case, human presence might be totally banned and exploration only be carried out using carefully sterilized robots and telemanipulators.

If no life is found in a particular region it could be re-assigned to the first category.

An interesting alternative is to combine the strategy of categorising the planetary surface as described above, with that of landing all missions in the same spot. The whole surface of the planet could then be studied by teleoperators, thereby imposing human control over an outpost located at a single spot. This would reduce contamination to a minimum, while the very presence of humans would make teleoperation possible. An extreme version of this strategy is to keep humans in orbit or on a satellite of Mars and allow only teleoperators to reach the surface.

While plans for future exploration are proceeding a bioethics debate should begin regarding whether it is appropriate to import terrestrial life to Mars and whether terraforming operations might later be implemented.

4.8. Backward contamination

Backward contamination comprises contamination of the Earth's environment by biological material coming from space and, in the case under consideration here, from Mars. This problem was taken into account already at the time of the *Apollo* project in that the astronauts coming back from the Moon were quarantined. However, it was thereafter recognized that the Moon is a non-biological body so that these precautions appear with hindsight to have been overestimated.

With Mars things are different and a heated debate started as soon as a sample-return mission was considered. Here we have two extreme positions: one stating that there is absolutely no danger because any parasitic organism evolves to infect a well defined type of life. Further, it is pointed out that meteorites with a Martian origin have reached the Earth several times so that, if contamination were possible, it would have already occurred. The fact in this scenario that these meteorites caused no contamination implies that there is no danger.

The opposing view is that of the ICAMSR (International Committee Against Mars Sample Return), a body that, although not opposing sample return missions in principle, stipulates that the samples should not be brought directly to Earth but instead be left for a long quarantine on the ISS where they can be studied, and perhaps stored forever. The committee holds that Mars is almost certainly inhabited by microorganisms which, like all bacteria, are potentially dangerous to any form of life. Exchanges of biological materials between planets, even if they happen naturally by means of meteorites, are extremely dangerous. They further hold that many epidemic diseases are potentially caused by meteorites, comets and asteroids.

Between these two opposing positions, the majority of scientists and the NASA administration maintain that reasonable quarantine measures for all Martian specimens must be implemented and that even more strict quarantine procedures **will be necessary when astronauts return to Earth from Mars.** These measures will obviously be required in particular in the circumstance that life is discovered during a human mission to Mars. The very fact that the search for life is presently one of the main scientific goals of Mars exploration guarantees that all Martian specimens will be studied in depth from this point of view and there is thus no chance that life forms which can interact with terrestrial life will go undetected.

It must be noted that the Mars surface environment is so harsh that if biological material is eventually found in protected places, it is quite unlikely to survive when exposed at the surface.

Usually not included in backward contamination is non biological matter. The surface of Mars is rich in chemicals hostile to all organic matter, and in fine dust which is harmful to both humans and machinery. These issues must be considered in designing the mission and in developing machinery to work on the planet's surface.

Organic compounds such as prions, although not living may be pathogenic. On Earth they are biologically produced but we do not know whether they might be the result of non-biological reactions. Even if the surface of Mars is found to be free of such substances, care must be exerted to prevent back contamination of this type.

Chapter 5.
HUMAN ISSUES

Long-duration crewed space missions pose issues and questions about crew safety and well-being that are critical for mission success. These issues center around three stressors: low gravity, radiation, and psychological and cultural factors. Such stressors may negatively affect human physiology, cognition, and individual and crew psychosocial status. In the following text, these issues and possible countermeasures for dealing with them will be discussed with reference to a human Mars expedition.

5.1. Physiological issues due to low gravity

5.1.1. Effects of low gravity on human physiology

Many, but not all, of the human physiological responses to spaceflight are functions of "g-transitions" and occur during and after dynamic phases of flight in which acceleration loads are acutely present and static gravity levels are changed chronically [1]. These include: launch from Earth and subsequent Mars transit trajectory; deceleration to capture Mars orbit or to land onto its surface; launch from Mars into orbit around the planet; acceleration to an Earth-return trajectory; and deceleration during final planetary re-entry. In between these dynamic phases, intrinsic biological mechanisms restore and maintain homeostasis as appropriate to the persisting environmental parameters. Thus, we expect the sensorimotor function to change most quickly in association with these dynamic flight phases and to reach homeostasis within a few days or a week. Body fluid redistribution, and the cardiovascular function which it drives, will be stimulated almost as quickly, but it requires a few weeks to accommodate. Red blood cell mass tracks body fluid volume with a delay of weeks to months. Muscle mass and bone integrity do respond to changes in physical workload associated with gravity levels, but very slowly – over periods of up to months. Their decrements in weightlessness may be obviated by rigorous physical exercise, as is being demonstrated currently on the ISS.

The relatively low Mars surface gravity (slightly more than 1/3 that of Earth) should not pose the same threat of orthostatic intolerance or bone fracture as ISS astronauts experience on return to Earth [2]. Unfortunately, it cannot be assumed that this low gravity will be useful in reversing the losses of sensorimotor, cardiovascular, and musculoskeletal capacity that occurred in transit. The astronauts' bodies will accommodate to their working and living environment homeostatically, but in the absence of definitive insight into the benefits of partial gravity, the best we might hope for is an interruption in further decrements for the duration of the surface stay.

Other body functions and organ systems are not clearly responsive to the forces of dynamic flight. Immune function may be dominated by the isolation of the astronauts within the closed spacecraft environment and may decrease as the mission progresses, leaving the astronauts at greater risk with each subsequent physical or environmental stressor. Exposure to the radiation environment of deep space (see below) will require shielding to minimize the potential of continuous disruption of chromosomal regulation of cellular health and proteomics, which may be repaired during the mission but may increase the lifetime risk of cancer and degenerative disease. Finally, the combination of all the stressors—physical forces, weightlessness, isolation, radiation, and limited possibilities for direct human interactions—will affect the behavioral health and capacity for performance of the astronauts in ways that are not yet well understood, given the relatively short duration of spaceflights to date (as compared with a Mars mission) and the as yet unknown human capacity for compensation.

Despite our current and improving understanding of human health in spaceflight, we should expect biomedical surprises during Mars missions. Recent experience on the ISS has revealed a deficit in visual function, perhaps related to months of exposure to the cephalic body fluid redistribution of weightlessness and possibly also involving increased intracranial pressure. Such visual function changes were long hypothesized and even documented in earlier shorter spaceflights, but their significance was underestimated until a few individuals experienced operational impacts when carrying out ISS work. We can be confident that this problem will be resolved, but it stands as a reminder that sometimes we actually know less about known problems than we think we do. A thousand day Mars flight presents abundant opportunities for experiencing surprises.

5.1.2. Suggested countermeasures.

As assessed by the NASA Human Research Program (HRP), many human health risks currently are or may be unacceptable, but they are amenable to reduction through research that provides evidence for mitigation. Operational risks associated with spacecraft mechanical safety may be greater than biological human health risks, but they typically occur during discrete and relatively brief mission phases. Human health risks may increase with exposure time and may persist well beyond the end of the mission.

Only a subset of all possible human health risks can be ameliorated. Conscious programmatic decisions will determine which risks ultimately remain unmitigated. Those decisions will not necessarily be "yea or nay" for each risk and may in fact constitute the inevitable result of constraints on the human research effort for unrelated reasons.

Risk reduction can be produced by minimizing exposure to risky environments. While simply being in space is risky, the threat correlates with the time spent in space, so that reducing transit times between planets would dramatically reduce exposure to the risk environment. The most elegant solution for many life sciences risks in space exploration may be outside the realm of space life sciences research, such as improved propulsion capabilities to shorten transit times by a large fraction, perhaps fifty percent. Such propulsion improvements are in early development, but they will require much larger power supplies than are currently available, thereby adding another pacing element to their realization.

Another risk reduction strategy would be to remove weightlessness as a major contributor to the physiological changes observed in spaceflight. Rotating all or a large portion of the transit vehicle could produce pseudo-gravity or "artificial gravity." Engineering considerations dominate this discussion: how to design a vehicle that can withstand the mechanical stresses of rotation without exceeding the propulsion system's capability to manoeuvre; ways to manoeuvre a rotating spacecraft; and whether or not to provide duplicate internal systems (such as hygiene and life support systems) to allow mission continuation in weightlessness if the rotation must be discontinued.

A point about artificial gravity is the presence of Coriolis acceleration due to spacecraft rotation: how much is this likely to cause discomfort to the crew? Another biomedical consideration of prime importance would be: how much gravity is enough? Even a half-century into the space age, definitive research has not yet been done to determine the long-term benefits of fractional gravity (between one-g at the Earth's surface and zero-g in orbital spaceflight). Any current postulations of adequate gravity levels less than one-g are the result of other considerations, such as engineering expediency, but not of rigorous biomedical inquiry.

5.2. Physiological issues due to radiation

5.2.1. Introduction.

This section deals with energetic particle radiation that can cause acute and chronic health effects in biological systems depending on: the magnitude of the radiation absorbed, the species of the radiation, the dose rate, the tissues affected, and the individual irradiated. The radiation encountered in deep space originates from Galactic Cosmic Radiation (GCR) and Solar energetic particle events. Doses due to galactic cosmic radiation can be 2-3 times higher at solar minimum than at solar maximum [3].

Hard spectrum solar energetic particles pose a major radiation hazard to crews beyond LEO. They potentially occur at any cycle phase and cannot presently be predicted [4].

5.2.2. Event modelling.

Galactic cosmic radiation exposure in free space during the minimum of Solar Cycle 23 was estimated using the European ***Crème2009*** model (employing the GEANT 4 code) and compared with published results obtained using the Langley ***HZETRN*** code ([5] and references therein). The results obtained using the two models were in reasonable agreement given inherent differences in the methodology adopted (those obtained using Crème2009 exceeded those derived using HZETRN by ~ 10%). It was demonstrated in the study that the dose incurred during 400 days in deep space (representative of a Cruise Phase to/from Mars) due to galactic cosmic radiation is hazardous in that the career dose limit values adopted by NASA (see below) for some of their space personnel would already be approached in the course of such a journey. This hazard would be further increased if galactic cosmic radiation levels present prior to

1957 were to return. The occurrence of a hard spectrum solar energetic particle event during the Cruise would constitute a serious superimposed health hazard.

The European Space Agency commissioned ‘***Mars Energetic Radiation Environment Models***’ (MEREM) to allow an assessment to be made of the environment potentially present at the Martian surface under different interplanetary conditions. An analysis made using MEREM for the cases of both solar maximum and solar minimum indicated that the radiation doses potentially incurred due to particles reaching the Martian surface at three selected landing sites were dependent on [6]:

- The epoch of the input galactic cosmic radiation.
- The fluxes of impinging solar energetic particles.
- The presence of good magnetic conductivity.
- The level of hydration and composition of the uppermost layer at the site considered.

Further, results obtained concerning surface dose levels, together with complementary results derived using HZETRN, indicated that the atmosphere of Mars would provide sufficient shielding to maintain the surface dosage due to GCR below the exposure limits adopted by NASA for their astronauts over a 30-day stay. The effect of the occurrence of hard (Carrington type) solar energetic particles during a human sojourn on the Martian surface was estimated to be sufficient to result in organ exposures in excess of NASA’s current permitted exposure limits [7].

There is now available a plethora of models constructed in different countries that predict solar energetic particle arrivals at Mars. These are based on different philosophies and must be merged in order to arrive at globally agreed upon perceptions concerning the differences between them [8]. Ground truth regarding model predictions can meanwhile be investigated using rovers and anthropomorphic phantoms, e.g., reports of data recorded at the Martian surface aboard the Curiosity Rover [9, 10]. Comprehensive data of this kind that is understood in depth should be available before any human mission is launched.

5.2.3. Astronaut exposure.

Permissible Exposure Limits (PELs) for crewmembers on deep space missions are chosen to prevent the taking of in-flight radiation risks deemed to be prejudicial to mission success, while also limiting chronic risks to acceptable levels based on legal, ethical, and financial considerations. Exposures are required to be ‘*As Low As Is Reasonably Achievable*’ (ALARA) to ensure that astronauts do not approach their assigned radiation limits while in flight. The application of ALARA dictates that measures are taken during the design and operational phases of the spacecraft to manage and limit crew exposures to ionizing radiation. Career exposure to radiation is estimated by NASA for individual missions and crews so as not to exceed a 3% Risk of Exposure Induced Death (REID) from fatal cancers. An ancillary requirement is that this risk limit is not exceeded at a 95% confidence level, using a statistical assessment of the uncertainties inherent in the risk projection calculations employed [11, 12].

Early radiation effects in humans are generally related to the loss of a fraction of cells that exceeds the threshold for impairment of function in a tissue. These effects are called “deterministic” because the statistical fluctuations in the number of affected cells are very small compared with the number of cells required to reach the threshold value.

The appropriate assignment of dose limits can ensure that early effects will not be experienced by the crew during a particular mission. Late effects can result from changes in a very small number of cells within which statistical fluctuations can be large and some level of risk is incurred even at low doses. These are referred to as "stochastic" effects. The relationships between radiation exposure, dose, and risk is age and gender specific due to: latency effects; differences in sensitivity between tissue types; and differences in the average life spans between genders. Prior crew exposure is also a relevant factor, and cumulative REID over several missions is considered when setting mission design requirements to ensure that the personal career PELs of individual crewmembers are not exceeded.

5.2.4 Dose Limits

At the present time, no space agency has assigned career dose limits for human personnel voyaging Beyond Low Earth Orbit (BLEO) and limits published thus far in the literature refer only to the Low Earth Orbit (LEO) environment. NASA specifies Short-term as well as Career Dose Limits for its astronauts. Short-term Limits are set to prevent the occurrence of clinically significant non-cancer health effects. In this regard a probability of 10^{-3} is deemed to be a practical limit for the risks that occur above a selected threshold dose. Career Dose Limits are intended to constrain the increased risk of contracting cancer incurred by members of the astronaut profession to an acceptable level.

With regard to short term and annual exposure, all the space agencies associated with the ISS have adopted a consensus limit for LEO. When defining Career Limits, certain agencies employ different constraints (for further details see [8]).

5.2.5. Microgravity conditions.

It is presently understood that biological results obtained under terrestrial conditions cannot be truly representative of what occurs in the space environment. In this regard, a variety of fluid redistribution effects and hormonal responses occur in microgravity which may influence cellular damage induction and repair systems, either directly or by controlling the state of oxygenation and the hydration of tissues. Indirect modification of circadian rhythms may also be involved. Overall, there are indications that the higher pro-oxidant state to which the human body adopts in micro-gravity may be part of a phase within which the deleterious action of ionizing radiation is mediated on a molecular, cellular and tissue level [13]. Extensive studies in this regard are in progress as well as investigations into the potential use by crew members of prophylactic radio-protective drugs and of possibilities to control phytochemical anti-oxidants in the human body through dietary choices [14].

5.2.6. Knowledge Gaps

A study mounted within the IAA (SG 3.19) which focused on investigating career dose limits for astronauts in LEO and the outlook for BLEO, identified several significant gaps in presently available scientific and technical knowledge relevant to supporting the safe implementation of human missions in deep space. Based on these gaps, recommendations were made to mount various international scientific and engineering based studies aimed at ameliorating the lacunae identified. Accounts of these recommendations and of the inherent problems concerned are contained in [8, 15]. Pending suitable advances in the enabling technologies required which include, among others: fast propulsion; customized spacecraft design and shielding; improved insights into the response of the human body to irradiation incurred under microgravity conditions; and advances in methodologies for reliably predicting solar energetic particle development and propagation, the best option presently available for improving the safety of humans in space is suggested to be effective on-board risk management based on the availability of reliable updating knowledge of the changing space environment [15].

5.3. Cognitive issues

5.3.1. The effects of space travel on cognition.

As mentioned above, extended periods of low gravity and radiation exert deleterious effects on the human body. To these physical factors we have to add the special environment characteristics of a spaceship for a mission of this kind: isolation, noise, space limitation, etc. These factors may have a negative effect on cognitive performance, underlying neurological structures, and brain mechanisms. Theoretically, the mental performance of astronauts may suffer from both direct effects on brain mechanisms by microgravity-induced neurophysiological changes and indirect effects related to stress-induced alterations of the attentional state [16]. Previous work has shown that various psychomotor and neurocognitive functions are degraded during spaceflight, among them central postural functions [17, 18], the speed [19-21] and accuracy [18, 22] of aimed movements, internal timekeeping [23], attentional processes [24], limb position sense [25, 26] and the central management of concurrent tasks [27].

There are several reasons to perform cognitive activities in space (Figure 5.1). These include clinical assessment, readiness to perform critical operations (e.g., transfer activities while in Mars orbit, space walks, landing, surface exploration, etc.), and research. The crew has to be prepared in case of any possible situation that may affect cognitive performance, and hence mission success or survival. Special environmental characteristics of long-duration space missions may affect performance, and some of these are also present in Earth analogues and space simulations. Cognitive aspects to be monitored in an expedition to Mars should include: attention, language, memory, learning, reasoning, and perception. Several cognitive tests and batteries such as MINICOG or the AGARD have been used in space missions and simulations, and the Spaceflight Cognitive Assessment Tool for Windows (WinSCAT) is the current standard for this type of assessment on the ISS. During interplanetary space missions, such as an expedition to Mars, crewmembers will experience increased autonomy. Moreover, there will be a time delay in communication between the crew and mission control back on Earth. Despite the utilization of self-administration assessment tools such as WinSCAT or the more recent Psychomotor Vigilance Test (PVT) [28] to detect risk factors in the cognitive and neurobehavioral domains, more integrated assessment and monitoring tools need to be developed and built into the on-board systems that can measure changes in different psychological areas, such as cognitive, psychosocial, and emotional/mood, as well as psychophysiological areas (e.g., pulse rate, EEG).

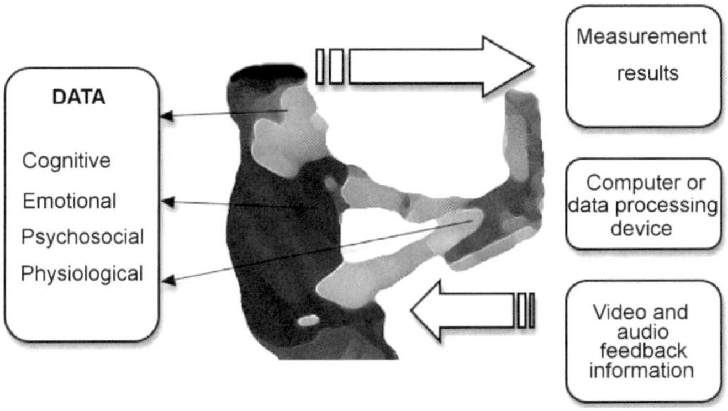

Figure 5.1. Integrated system for performance monitoring.

Similar to the risks and unknowns faced by polar explorers, astronauts embarking on an interplanetary mission to Mars will perform at their best because of deeply embedded human urges to explore.

5.3.2. Suggested countermeasures.

The type of integrated system described above may help to provide and apply the best suitable countermeasures possible. Monitoring and detection measures can be based on pattern recognition techniques that will record the facial expressions of individual crewmembers and analyze these in relation to surrounding conditions and actions of other members of the crew. The technology should be able to interpret the body language of individual crewmembers, their posture, their interactions, and their facial expressions. Through collation and analysis of facial and body language data, together with physiological markers such as pulse rate, it may be possible to detect positive and negative interaction patterns between crewmembers.

This technology will also be able to identify the location of each person, his or her proximity to others, their frequency of interaction, time spent on work and at leisure, and correlate these factors with performance and emotional state. Furthermore, together with data collected from facial and voice recognition, cognitive performance, and by biosensors, we can gain insight into patterns of performance for each subject and for the overall team. Two important aspects to be monitored in the cognitive domain for long-term missions are reaction time and accuracy. Reaction times vary significantly in different circumstances, such as neurological disease, brain injury, and under stress or fatigue. Computerized systems allow crewmembers to track reaction times and accuracy levels in the tests they perform to detect clinically significant variations in these indices. They can also provide insight on the crew's coping and resolving strategies.

5.4. Psychological and cultural issues

5.4.1. What do we know from on-orbit missions?

There have been several research studies involving astronauts and cosmonauts that have given us information about important psychological, interpersonal, and cultural issues that affect space crewmembers. For instance, two NASA-funded international studies of psychological and interpersonal issues during on-orbit missions to the Mir and the ISS were conducted [29-33]. A total of 30 crewmembers and 186 mission control personnel were studied. Subjects completed a weekly questionnaire that included items from a number of valid, well-known measures that assess mood and group dynamics. There was significant evidence for the displacement of tension and negative emotions from the crewmembers to mission control personnel. The support role of the leader was significantly and positively related to group cohesion among crewmembers, and both the task and support roles of the leader were significantly related to cohesion among people in mission control. Russians reported greater language flexibility than Americans. Americans scored higher on a measure of work pressure than Russians, but Russians reported higher levels of tension on the ISS than Americans. There were no significant changes in levels of emotion and group interpersonal climate over time and no general indication of the so-called "third-quarter phenomenon", i.e. a finding that some crewmembers in isolated and confined environments experience depression and other negative emotions after the half-way point of their mission [34].

A study used an analysis of speech patterns and a measure of subjective attitudes and personal values to study on-orbit space crews and people working in space analogue environments [35-39]. It was found that, over time, these isolated groups showed decreases in the scope and content of their communications and a filtering in what they said to outside personnel, which was termed psychological closing. Crewmembers interacted less with some mission control personnel than others, perceiving them as opponents. This tendency of some crewmembers to become more egocentric was called autonomization [35, 36]. The research team also found that crewmembers became more cohesive by spending time together [37], and that the presence of subgroups and outliers (e.g., scapegoats) negatively affected group cohesion [38]. In a study of twelve ISS cosmonauts it was found that personal values generally remained stable, with those related to the fulfilment of professional activities and good social relationships being rated most highly [39].

An examination of potentially disruptive cultural issues affecting space missions was carried out using a survey of 75 astronauts and cosmonauts and 106 mission control personnel [40]. The subjects rated coordination difficulties between the different space organizations involved with the missions as the biggest problem. Again, a study surveyed eleven cosmonauts regarding their opinions of possible psychological and interpersonal problems that might occur during a Mars expedition [41]. The following factors were found to be rated highly: isolation and monotony, distance-related communication delays with the Earth, leadership issues, differences in space agency management styles, and cultural misunderstandings within the international crew. A survey of 576 employees of the European Space Agency identified a link between cultural diversity and the ability of people to interact with one another [42]. Especially important were factors related to leadership and decision-making. Finally, a content analysis was made of personal journals belonging to ten ISS astronauts [43]. These were oriented around a number of issues that had behavioural implications and showed that 88% of the entries dealt with the following categories: Work, Outside Communications, Adjustment, Group Interaction, Recreation/Leisure, Equipment, Events, Organization/Management, Sleep, and Food. In general, the crewmembers reported that their life in space was not as difficult as they had expected it to be prior to launch, despite a 20 percent increase in interpersonal problems during the second half of the missions.

5.4.2. The Mars 500 Program

From June 2010 to November 2011, a unique ground-based space analogue mission took place that was called the Mars 500 Program [44]. It was designed to simulate a 520-day round trip expedition to Mars, including periods of time where the crew functioned under high autonomy conditions with communication delays with outside monitoring personnel in mission control. Six men were confined in a simulator that was located at the Institute for Biomedical Problems in Moscow.

Several psychosocial studies were conducted during the 520-day mission and changes in crewmember time perception, evidence for the displacement of crew tension to mission control, and decreases in crewmember needs and requests during high autonomy were found, which suggested that they had adapted to this condition [48]. It was reported that the crew exhibited increased homogeneity in values and more reluctance to express negative interpersonal feelings over time, which suggested a tendency toward "groupthink" [49]. It was shown that the crewmembers experienced increased feelings of loneliness and perceived lower support from colleagues over time, which had a negative effect on cognitive adaptation [50]. Wrist actigraphy, the psychomotor vigilance test, and various subjective measures were used to study the crew and a number of individual differences in terms of sleep patterns, mood, and conflicts with mission control were found [51]. Finally, an evaluation of fixed video recordings of crew behaviour during breakfasts identified variations in personal actions, visual interactions, and facial expressions, but a general decrease in group collective time from the outbound to the return phase of the simulated fight to Mars [52].

During a 105-day pilot study in 2009 that preceded this mission, a study was made of the moods and group interactions of a six-man Russian-European crew and the relationships of this crew with outside mission control personnel [45, 46]. Employing measures similar to those used in their on-orbit work [26-33], the research team concluded that high work autonomy (where the crewmembers planned their own schedules) was well-appreciated by the crew, mission goals were accomplished, and there were no adverse effects, which echoed recent positive autonomy findings in other space analogue settings [47]. During the high autonomy period, crewmember mood and self-direction were reported to be improved, but mission control personnel reported more anxiety and work role confusion. Despite scoring lower in work pressure overall, the Russian crewmembers reported a greater rise in work pressure from low to high autonomy than the European participants.

5.4.3. Space psychiatry and salutogenesis

A number of psychiatric problems have been reported during on-orbit space missions [26]. Most common are adjustment reactions to the novelty of space. These largely consist of transient anxiety or depression. Psychosomatic reactions also have been reported. Asthenization, a syndrome consisting of fatigue, irritability, emotional lability, and attention and concentration difficulties, has been reported to occur commonly in cosmonauts by Russian flight surgeons [53]. Problems related to major mood and thought disorders (e.g., manic-depression, schizophrenia) have not been reported during space missions, probably because crewmembers have been screened psychiatrically for predispositions to these psychotic conditions before launch. Post-mission personality changes and emotional problems have affected some returning space travellers. These have included depression, anxiety, alcohol abuse, and marital readjustment difficulties that in some cases have necessitated the use of psychotherapy and psychoactive medications [29].

Isolated and confined environments can also produce positive experiences [54]. A survey of thirty-nine astronauts and cosmonauts found that all of the respondents reported positive changes as a result of flying in space [55]. One subscale especially stood out: 'Perceptions of Earth'. One of the items in this subscale which dealt with gaining a stronger appreciation of the Earth's beauty had the highest mean change score. Extended pioneering research was begun in the early 1990s on the salutogenic (or growth-enhancing) aspects of space travel, In this regard an analysis was made of the published memoirs of 125 space travelers [56]. After returning from space, the subjects reported higher levels on categories of Universalism (i.e., greater appreciation for other people and nature), Spirituality, and Power. Russian space travellers scored higher in Achievement and Universalism and lower in Enjoyment than Americans. Overall, these results suggest that traveling in space is a positive and growth-enhancing experience for many of its participants.

5.4.4. Suggested countermeasures

Despite their relevance for future Earth orbit and lunar missions, the above findings may have limited general applicability to long-distance, multi-year expeditions such as a mission to Mars, where new stressors will occur related to autonomy, two-way communication delays of up to 44 minutes, and extreme loneliness due to perceiving the Earth as an insignificant dot in the heavens (the so-called Earth-out-of-view phenomenon) [57].

There are several countermeasures that can be implemented to help ameliorate the impact of the above issues on the crew of a Mars expedition in terms of selection, pre-launch training, mission monitoring and support, and post-return adaptation. Crewmembers should be selected who are sensitive to psychological issues, comfortable working alone on a project, and able to interact socially with their teammates when this is appropriate. Commanders should be chosen who have both task-oriented and supportive leadership skills. Pre-launch training should include both didactic and experiential sessions that deal with important psychological, interpersonal, and cultural issues that may occur, as well as learning to work effectively under high autonomy conditions. Conjoint training involving both crewmembers and key mission control personnel should take place and consider issues related to displacement and possible crew-ground miscommunication [29]. Computer-based refresher courses echoing some of this training should occur during the mission itself. Crews should plan time during the mission to identify and deal with stressful personal and interpersonal issues before they fester and become problematic. Crewmembers should develop ways of communicating under time-delayed conditions with people on Earth, such as adding anticipatory questions at the end of email messages that minimize the need for repeated back-and-forth communications. The Earth-out-of-view phenomenon could be addressed by providing an on-board telescope. The crewmembers should know that their families at home are being supported during the mission through formal or informal group activities. Readaptation debriefings and private time together need to be scheduled post-return in order to help the crewmembers and their families readjust to each other and deal with fame and glory issues resulting from a highly-publicized space expedition.

Chapter 6
THE SPACE TRANSPORTATION SYSTEM

6.1. Introduction

Defining the architecture of the mission is a well-known problem. The two main difficulties are:

- The number of possible options.
- Defining scenario criteria
- Eventually to assessing options in each mission phase.

These difficulties are probably the main reasons for the problems with the many Mars mission studies presented so far (see Chapter 2). For instance the feasibility of the last NASA mission architecture, the most detailed analysis on this subject, is still uncertain (DRA5 2009).

Uncertainties and risks can be reduced by means of long and expensive studies of every critical segment of the mission. For complex systems, uncertainties will remain high and the only solution might be to increase each segment's TRL with intermediate missions, which would drive the costs even higher. **Risk decrease is therefore closely linked to cost increase**. Thus decision making must include feasibility, risks, but also mission cost and roadmap sustainability. Several studies on methodological issues linked to risk quantification and risk analysis for decision- and policy-makers have been performed [23, 27].

In order to reduce the risks to an acceptable level (i.e. human rated space systems must be flight proven) more work is needed, and this will raise the overall mission cost.

The total cost of the first mission will include:

- Development of new facilities, space modules and systems (tests and qualification included)
- The cost of preparatory missions required to increase the maturity level and reduce the risks of the transportation system (e.g., maturity of Martian entry, descent, and landing systems).

It is a key fact that any preparatory mission to develop ancillary systems not directly related to the Mars mission will raise the cost. All in all, the cost of the first mission will probably be several times higher than the costs of follow-on missions. In fact, a program roadmap lasting decades and mobilizing most resources of space activities without immediate results could become a showstopper.

As a consequence, principles driving the design of architectures can be phrased as:

- The mission must be as simple as possible.
- Whenever cost effective and possible, simple technologies with high technology readiness levels (TRL) must be used.

- The organization must be simple and efficient. Typically, a mission program involving many or complex spacecraft with complex LEO assemblies should be avoided.

These principles may be contradictory. It is well known that the use of new and more efficient types of propulsion may simplify scenario and organization, but at the expense of the lower TRL and higher development costs to qualify the new systems.

A synthesis of the most critical options is presented below:

- Crew size
- Launcher and LEO strategy
- Spacecraft configuration
- Interplanetary trajectory
- Interplanetary propulsion system
- Mars orbit insertion
- Descent vehicles and Entry Descent and Landing (EDL) strategy
- Mars orbit to Earth orbit strategy
- In Situ Propellant Production (ISPP)
- Overall redundancy and multiple mission strategy

In the following sections each parameter is examined, a cross impact analysis is presented and the main options are highlighted.

6.2. Crew size

Crew size significantly impacts spacecraft mass and volume budgets, thus impacting LEO assembly, the number of launches from Earth, EDL options (differing in TRL, complexity and risks), feasibility of aerocapture, the Earth return segment and the interplanetary propulsion system [24]. The deep space habitat mass is a function of crew size, see Table 6.1 [20]; when the crew is doubled from 3 to 6, the habitat mass increases 1.5 times. This has a direct impact on many systems, architecture complexity and especially cost (LEO orbiting costs are in the 7,000 – 20,000 $/kg range).

Duration	**600 days**	**800 days**	**1000 days**
2 astronauts	19.7	22.2	24.9
3 astronauts	23.5	26.8	30.1
4 astronauts	27.1	31.1	35.2
6 astronauts	34.0	39.7	45.3

Table 6.1 Mass in metric tons of the habitable module as a function of crew size and mission duration [20]

Cost and risks suggest to keep the number of astronauts a strict minimum. However, a small crew brings a narrow range of skills [17, 19]. To compensate, candidate astronauts can train to reach the know-how required to fill the gaps. Also, many tasks can be automated [7] or can benefit from the expertise and support from mission control [11].

If the learning skills can be managed, optimizing crew size depends on minimizing risk. For other mission systems the "2-fault tolerance", the minimum requirement for human rated space systems, is recommended. In the 2004 ESA reference Mars mission it is suggested to follow this principle, and to send a crew of three astronauts [8]. Three astronauts is not optimal for reasons linked with human factors, which suggest that five or six would be better, despite the increasing mission cost. According to psychology and physiology experts, a three-astronaut crew might be acceptable provided they are properly selected and trained, and as such is recommended in this Study.

If adding one or two astronauts does not impact much the complexity and cost of the mission, this option should be seriously considered. Providing that it is practical, a simple but very costly way to send more astronauts to Mars is to duplicate the mission, i.e. to send two spacecrafts so that two small crews would be always close to each other during the interplanetary segment, and after landing would share the same habitat. This would provide redundancy in a crisis.

Recommendations

- Plan a minimum crew of 3
- Carry on parametric assessments of other options.

6.3. Launcher and LEO strategies

6.3.1. Assumptions

As suggested in most Mars mission architectures, it is assumed here that **at least 3 interplanetary vehicles** (two cargos and one crewed) are sent to Mars. They will carry the Mars lander/ascent vehicle, the interplanetary and Mars surface habitat, and the propulsion system for the return. It is also assumed that a **total IMLEO is in the range 500 to 1000 tons**.

6.3.2. Transportation to LEO

At $7-20k per kilogram, the cost of transportation to LEO is the largest fraction of the recurring cost of a Mars Mission. Using the highest values, a total launch cost of $20 billion is obtained. Super heavy lift launchers with a 100 ton LEO capability are not available but have been mastered in the past (Saturn V, Energia) and will be built (e.g., the NASA SLS Block 2 has 130 t lift capability; the Long March 9, developed by the Chinese Space Agency, is similar).

A major problem with heavy launchers is the launching rate. Once the decision to launch is taken construction, assembly, tests and launch preparation require about one year, and monopolize industrial assets. In addition, in industry it is difficult to maintain an unsteady ('stop and go') pace over a long period of time. Ideally, there would be one construction and launch per year, very optimistically two, but this also depends on the type of mission. If the heavy launcher is only used for the Mars mission, such more realistic production rate of one every two years is industrially unappealing.

These are key constraints and must be considered in planning mission architecture and in preparatory phases.

6.3.3. Trans-Mars injection requirements using Hohmann orbits

In the following, a single notional vehicle and trajectory are assumed to analyze propulsion requirements. Starting from a 400 km LEO, the Mars transfer vehicle must perform a ΔV of 3700 m/s for a 200 days coasting phase trajectory to Mars with one or two mid-course man oeuvres in between [15]. Higher ΔV at injection would shorten the transfer time but would also result in higher velocities at arrival, which increases cost (in ΔV) and complexity of the Mars orbit injection.

In order to inject a 40-45 ton payload (the assumed mass of the interplanetary vehicle) from LEO to a Mars transfer trajectory by means of LOX/LH_2 chemical propulsion, a launcher with 130 tons LEO capability is required [16]. This is probably sufficient if the crew is reduced to 2-3 astronauts or 2 crews of 2 [22]. Adding a Mars ascent vehicle, a surface habitat and the propellant for the return vehicle would roughly double the LEO mass. Therefore, a superheavy – and unaffordable – launcher of roughtly 300-ton LEO capability would be required to accelerate the 100 ton payload to the trans-Mars injection (TMI) orbit. LEO assembly of the interplanetary vehicle is therefore unavoidable. The alternative is another type of propulsion system: still assuming 100 tons for the TMI vehicle, the mass lifted to LEO reduces to 240 tons using nuclear thermal propulsion and even less using electric propulsion. This key point is discussed in the section comparing different interplanetary propulsion systems.

6.3.4. Assembly of several modules to build a Mars transfer vehicle

A reference assembly orbit is a 400 km circular parking orbit. Here assembly means the autonomous and automatic rendezvous and docking (like the ATV for the ISS) of several propulsion modules and the payload module to a train [10].

The simplest solution is to assemble the Mars injection propulsion stage in LEO with a 65-ton payload module. A 130 ton chemical stage can typically boost a 65 ton payload module to Mars, so the initial mass of this Mars transfer vehicle would be 195 tons. The advantage of this method compared to direct TMI injection is a reduction of the launcher size and a possible increase of the departure rate (several payloads can be lifted to LEO before launch period and wait in LEO for rendezvous). NASA SLS Block 2 launcher has just a 130 ton LEO capability. If zero boil-off technologies are not mastered, only one payload per opportunity can be sent to Mars from a single launch pad. Eventually, if the waiting time in LEO is not too long, boil-off losses of cryogenic propellant might be acceptable but this remains an issue.

The main advantage of the LEO assembly option is flexibility. The relative independence of the launch rate from the Mars transfer perios means Earth launches can be spread out over time. As the injected payload mass is not limited a fleet of several vehicles can, at a chosen opportunity, be sent to Mars simultaneously.

The main drawback of any LEO assembly is that, from time to time, a re-boost of the different modules is necessary to keep on the 400 km orbit, which results in a significant mass penalty. In the last NASA DRA5 architecture, the mass of the re-boost modules is greater than 100 tons [19].

Zero boil-off technologies are mandatory (*sine qua non* condition) with LH_2, including super insulation techniques and efficient cryo-coolers systems; the latter of which have not yet been mastered. Direct injection scenarios do not require zero-boil off technology and reduce mission complexity (no rendezvous and no docking), but are limited to a launch roughly every 2.1 years. Since the Mars launch period lasts only around four weeks, to orbit more than 130 ton requires simultaneous construction and liftoff of several heavy launchers. The impact on logistic, industrial capability and facilities – such as launching pads – would be dramatic. The direct injection strategy is more appropriate in the case of "split" architectures, which suggest, for instance, that the Mars ascent vehicle and the surface habitat be sent to Mars (or Mars orbit) separately and in advance.

Recommendations

The mass of the re-boost modules penalizes architectures based on giant interplanetary vehicles and long duration LEO constructions. It is recommended to

- Keep developing heavy launchers with 130 ton LEO capability
- Avoid LEO assembly or minimize LEO waiting time;
- Minimize the mass of interplanetary vehicles by splitting payload into modules sent to Mars separately;
- If possible, send only one vehicle to Mars per TMI period (split strategy).

6.4. Vehicles configurations

There are many options for the number and configuration of interplanetary vehicles, the number and configuration of the Mars landers, the pre-deployment strategy, and the configuration of the Mars Ascending Vehicle (MAV). A few examples include:

- A cargo vehicle that can be sent to Mars in advance with some assets that can be deployed on the surface before the manned vehicle with is sent to the red planet. This stategy is interesting because it provides the landing crew with a MAV that is ready and waiting for take-off in case of an emergency. Pre-deployment may however be very difficult without human intervention – some ISRU options might be penalized – and, according to NASA, the long waiting periods in orbit bring other risk issues. Note that split missions are not feasible unless precise landing/guidance can be ensured: the crew must walk to the cargo, for instance to deploy a rover. The need for guidance suggests orbiting a Mars Guidance, Navigation and Control (GN&C) satellite prior to any exploration.
- The same crewed interplanetary vehicle may be used for the outbound and inbound trips [19], or it may be different [34].
- The crewed interplanetary vehicle may also be a lander. The alternative is to land a separate descent vehicle that will dock in Mars orbit with the return spacecraft.
- The MAV may be integrated in the crewed descent vehicle and the same propulsion systems be used for landing and ascent. The other option is to send the MAV in a cargo to be pre-deployed.
- The MAV may include the main surface habitat [22].
- The MAV may be the Earth Return Vehicle (ERV) for a direct return [33].

- A pressurized rover may land as part of the surface habitat, or in a crewed cargo [26]. The surface habitat itself might be a pressurized vehicle.

These are only a sample of the many possible configurations.

Figure 6.1 shows a possible ERV configuration. The ERV can be split into two smaller vehicles that proceed to aerocapture independently and join in Mars orbit later. The first vehicle includes the main propulsion stage for Trans-Earth Injection (TEI) and the Earth re-entry capsule. The second includes the interplanetary habitat and a small service module that can be jettisoned after the junction with the propulsion stage. At the end of the stay on the surface, they are joined to the Mars ascent vehicle. If the surface habitat is small (small crew size), it might even be possible to integrate it in the MAV.

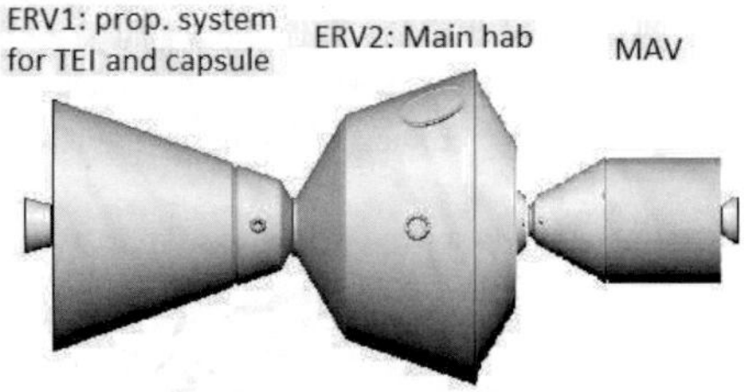

Figure 6.1: Possible configuration for the ERV.

Recommendations

Examine all options. An optimized configuration might reduce IMLEO more than a higher specific impulse.

6.5. Interplanetary trajectories

6.5.1. Missions without landing

Missions which do not include actual landing on Mars have been proposed several times. The simplest is a flyby, as the one proposed by Inspiration Mars. This approach is based on a free-return trajectory (in the 2018 launch opportunity. The ‘plan B’ option for the 2021 launch opportunity is based on a much longer Venus gravity assist trajectory) which takes about 500 days round-trip. Flyby missions will not be considered here, mainly because their scientific returns are quite low compared with crew time spent in space and the associated risks. A flyby mission was run in the *Apollo* program to test all the equipment, but the Moon is much closer than Mars, and today testing missions can be run without humans on board. While a private operator flyby mission might be justified, it may not in the frame of an international mission, even as a ‘dress rehearsal’.

Missions in which the crew remains in orbit around Mars or land on one of its satellites have been proposed several times, motivated by the dangers posed by the Mars environment. This consideration is now considered outdated. The advantages of robotic exploration controlled by human teleoperators in orbit around Mars (on a spacecraft or on a satellite) are:

- Reducing dangers of forward (and backward) contamination
- Reducing the risks and complexities of EDL (but aerobraking is needed anyway)
- Reducing the risks and complexities of ascent from Mars

The price to pay consists of:

- Reduced scientific efficiency and presumably output
- Almost no impact on future Mars colonization
- Increased risks due to longer stay in space, including psychological and physiological trauma due to radiation and microgravity
- Additional cost

A Mars orbital mission without landing might be considered to test equipment, (the EDL and MAV devices can be tested by downloading teleoperators and uploading samples from the planet) for future missions on the ground. They will not be dealt with any further in this study.

6.5.2. Conjunction / opposition mission

Owing to the relative positions of the planets, the two options are a conjunction (long stay) or an opposition (short stay) mission. The trade-off between the two options has been discussed in several papers [9, 12, 17, 19, 33]. A summary of the arguments is presented here.

The most economical – from an energy viewpoint – Mars mission is a long stay following an elliptical Hohmann (or close to Hohmann) trajectory taking about eight to nine months. The transit time can be reduced to five to six months with more propellant. The drawback is the duration of the stay: about 500 days, waiting for the two planets to be in orbital positions minimizing propellant consumed to return to Earth.

A short stay mission is the second option; the price is increased transit time and propellant consumption. Departure from Earth should generally be performed earlier in order to rendezvous with Mars a few weeks or months before the conjunction of the two planets and to be able to return to Earth. A Venus gravity assist may eventually help reducing propellant mass, but this would still be much greater than for the long stay option.

This trade-off is illustrated by the two trajectories in Figure 6.2. Time spent in space and on the planet is in Table 6.2, where numbers depend on the launch period. Journey duration depens on launch time, owing to the fact that the orbits of the two planets are elliptical and not exactly coplanar.

In terms of science, a short stay mission leaves little useful time to accomplish the mission goals, and while exposing the crew to a higher radiation dose and microgravity effects during the longer transit. For example, a Venus flyby would further increase the dangers due to radiation.

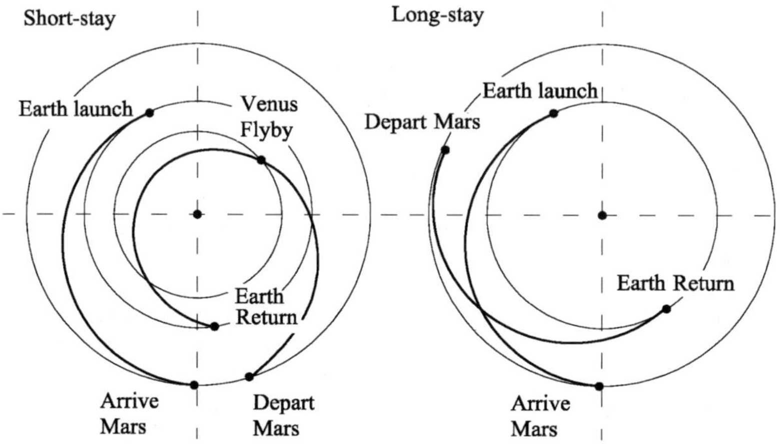

Figure 6.2. Sketch of the trajectories, with the Sun at the centre, for a short-stay (opposition) and a long-stay (conjunction) Mars missions.

	Flyby	Short-stay	Long-stay
	Days	Days	Days
Outbound	228	224	224
Stay	0	30	458
Inbound	273	291	237
Total	**501**	**545**	**919**
Total in space	501	515	461
travel/stay	∞	17.2	1

Table 6.2. Time spent in space and on the surface of Mars for the two missions considered in Figure 6. 2, compared with the time spent in space for a flyby mission.

The second option lengthens the mission but also increases time on Mars surface by a factor ten; transit time is reduced by roughly 10% and the spacecraft never gets closer to the Sun than in Earth's orbit.

Most studies comparing conjunction and opposition missions for different launch periods agree that conjunction (long stay) missions are preferred for the first mission, in order to reduce IMLEO [17, 19].

This conclusion holds for chemical and probably NTP systems. High power NEP systems enable faster missions, thus allowing for missions with intermediate stays. NEP-powered trajectories are no longer Hohmann, and thus are faster, for electric power in the 100 MW range [9]. Other mission schemes suggested are based on 'cyclers', large spacecraft traveling on trajectories tangent to those of Earth and Mars about the Sun. These schemes are not considered here, since they are insufficiently studied for a first human Mars mission. However, such orbits include interesting backup options such as "free return".

A different Mars mission proposal may be called ‘indefinite stay’. This is a one-way mission, justified on the ground that if the purpose of Mars missions is colonization, astronauts should remain on the planet for the rest of their lives. In this case one of the key mission objectives, returning to Earth, is replaced by that of surviving indefinitely on Mars. This approach is very controversial, and will not be discussed here any further.

6.5.2. Free return trajectories

In case of major problems during the outbound leg, a free-return trajectory would allow the crew to return to Earth without any major maneuvering [12, 31]. Several strategies exist to achieve this. The simplest is to choose an elliptic trajectory that returns to the departure point (the perihelion) after a period of exactly 2 years. Such trajectories go farther than the orbit of Mars. Mars rendezvous is feasible by choosing the right time for launch. If it is desired to abort the mission very early, it would just be required to avoid Mars or to provide a soft swing-by maneuver. Other free return options are possible [12], but a key factor is the length of ‘free return’: this may be excessive in an emergency.

6.5.3. Launch Periods

Launch periods to Mars are scarce and short. Launch opportunities happen only once every 780 days and rarely last more than 1 month. All the launch opportunities do not require the same propulsion effort (the ΔV), which is lower for launching about three months before a perihelic opposition. After 2018, the next perihelic opposition will occur in 2035. Considering low energy transfers from high Earth orbit, ΔVs are significantly different. This difference is less than 10% for trans-Mars injection from low Earth orbit [9]: a 3.4 km/s ΔV is required to reach Mars perihelion and 3.7 to reach its aphelion. For a 2 years free-return trajectory, the ΔV is 4.3 km/s from LEO.

Recommendations

- With chemical propulsion choose conjunction missions and long stays reducing the IMLEO.
- Consider free return trajectories as backup options in case of early mission abort.

6.6. Interplanetary propulsion systems

A strategic question raised in this study was the type of propulsion: chemical, nuclear (nuclear thermal or nuclear electric) and solar. After long discussions, there is still no consensus among members of this IAA study. The importance of this issue suggested summarizing propulsion options in Appendix C. The following three sections show examples of achitectures based oon NTP, chemical propulsion and electric propulsion, namely SEP. Sections from 6.6.4 to 6.6.6 give some reasons which may influence a choice between the various solutions, although no recommendation is provided.

6.6.1. Architectures based on nuclear thermal propulsion systems

This option has been studied since 1997 in great detail at NASA to design reference architectures. [2, 17, 19, 20]. The main features are:

- Crew of 6.
- Conjunction class (long stay).
- At least 9 launches (130 ton in LEO each) by a heavy launcher to assemble 3 interplanetary nuclear thermal rockets:
  - The first is a cargo deploying the MAV in advance
  - The second is a cargo carrying the surface habitat module from LEO to a Mars orbit. Both cargoes are launched before the crewed spacecraft – possibly in the preceding launch period.
  - The third carries crew and their deep space habitat during the interplanetary travel between Earth orbit, Mars, and back
- Aerocapture is performed by two cargoes but not by the crewed vehicle.
- On arriving in Mars orbit, the crewed vehicle rendezvous with the surface habitat module. This lands on the surface close to the MAV.
- Mars is explored during the 500-day stay.
- The MAV is used to lift from the surface to orbit, to rendezvous with the nuclear thermal rockets and to return to Earth.
- The Orion capsule attached to the crewed nuclear thermal rockets during the entire mission, re-enters Earth's atmosphere.
- The total IMLEO is around 900 tons.

An artist's drawing of the three nuclear thermal rockets is in Figure 6.3.

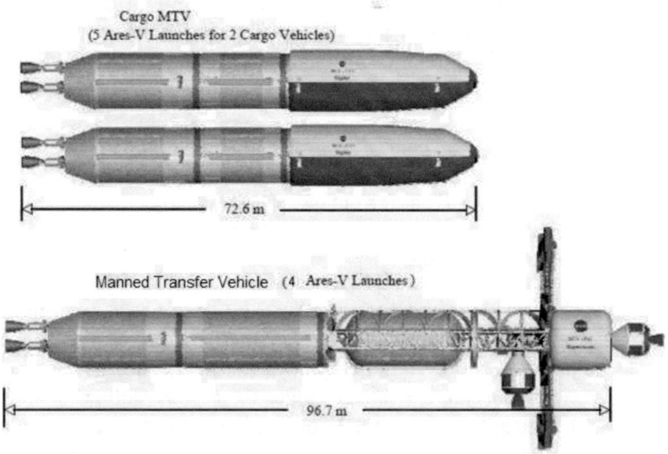

Figure 3: Illustration of the three NTR planned in NASA's DRA 5 (courtesy NASA).

6.6.2. Architectures based on chemical propulsion systems

Proposals are many [8, 22, 26, 33]. Some are overly complex; others rely on optimistic estimates of module size and mass. More so than with nuclear propulsion, optimistic assumptions and estimates impact not only propulsion, but the entirety of the mission. In addition, architectures saving IMLEO may not be robust enough, or provide enough margins of error in key mission phases.

Among architectures, the "2-4-2" concept, allegedly much simpler and economical in terms of IMLEO than the NASA mission [22], is described in some detail below. The key element of the 2-4-2 architecture is the use of aerocapture enabled by reduction of the ballistic coefficient. The ballistic coefficient is reduced by splitting the payload into two smaller and lighter landers, each carrying a crew of two. Aerocapture simplifies entry, descent, and landing phases. This strategy enables launching the Mars habitat back to Mars orbit when desired. Optimization helps to reduce IMLEO. Its main features are:

- The crew of 4 is split in two crews of 2.
- The mission is entirely duplicated.
- Mission is conjunction (long stay).
- six launches of a 100 tons LEO capability launcher are required to send six interplanetary vehicles to Mars.
- As the mass of each interplanetary vehicle is less than 40 tons, there is no need for LEO assembly.
- Aerocapture is performed by all vehicles with a single optimized aeroshell.
- Each cargo brings ISRU equipment to the surface.
- Each habitat is dual use: it is used in space for the outbound and inbound trips *and* also on the surface of Mars.
- Each return vehicle is sent to Mars orbit and is made of two elements: a wet propulsion system and a small capsule. The capsule has a heat shield and will be used only to re-enter Earth; it also stores consumables for the trans-Earth return segment. At the end of the stay on the Martian surface, the main habitat module is lightened, and ascends to orbit where rendezvous with the return vehicle.

6.6.3. SEP architectures

Several Solar Electric Propulsion (SEP) architectures have already been proposed [13, 25, 29]. The main concept is to employ large arrays of photovoltaic cells in LEO in order to supply continuous power to ion thrusters with small thrust but high Specific Impulse (ISP). In general, SEP architectures require four or five heavy payloads to send all modules needed to a Mars mission. With a tug supplying at least 300 kW, 50 tons of payload can typically be sent to Mars with a 40-50 ton SEP vehicle. The problem is the weak thrust provided by ion thrusters, in the order of 40mN/kW. More thrust can be obtained with larger solar panels but at the expense of the complexity of their deployment. 300 kW can reasonably be achieved without complex LEO assembly but higher power would pose problems. The implication is that it would take a very long time to reach high Earth orbits. Starting from LEO, provided a 300 kW solar tug is available, the thrust is so low that it takes several months for a 50 ton payload to escape Earth. As a consequence, SEP is appropriate for the transfer of cargo vehicles but less so for crewed vehicles. For crewed vehicles, 2 options are generally proposed:

- Chemical propulsion only (or nuclear thermal).
- Solar electric for an automated vehicle module to reach high Earth orbit, and chemical propulsion for a crewed capsule to perform a rendezvous with the uninhabited module. This option is usually preferred because it is often assumed that a crewed transport capability is necessary for other uses and will be soon available (e.g., the Orion vehicle).

An original architecture is proposed to illustrate SEP, see Figure 6.4.

Solutions based on SEP are not much different from those based on NEP: in both cases the thrust is low, and a long time is required to reach escape speed. A slightly smaller time period is needed to reach the TEI to the return orbit. Actually in both cases the duration of the journey depends only on a single parameter: the specific mass of the power generator, i.e. the ratio between the mass of the generator m and its power P, usually indicated with $\alpha = m/P$. If a low value of α can be obtained, low thrust interplanetary transfer can be quite short, even much shorter than those achievable with chemical and even nuclear thermal propulsion. At the state of our electric generator technology $\alpha = m/P$ is of order of 10 kg/kW, still inadequate for medium or fast interplanetary orbit times.

A problem specific to SEP is that α increases with increasing distance from the Sun (because the power produced by the solar array decreases). The deceleration phase at Mars and the injection in the trans-Earth trajectory at the return are therefore more problematic than in the case of NEP.

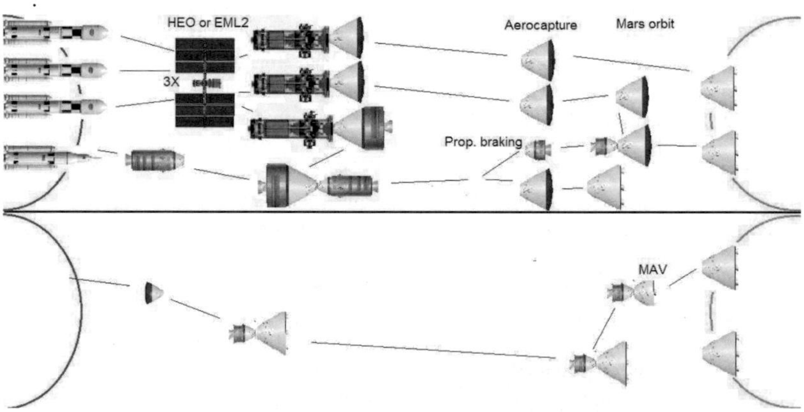

Figure 6.4: A possible SEP architecture.

The recent development of variable ISP plasma thrusters, like the Variable Specific Impulse Magnetoplasma Rocket (VASIMR) engine may allow SEP and especially NEP to become a reality also in case of large spacecraft, particularly if this engine will be developed for reboosting the ISS. Regarding SEP, the problems due to high α still remain. Regarding NEP, or combinations of NEP and NTP, insufficient analysis of missions exists to draw conclusions

6.6.4. Reasons for choosing nuclear propulsion

A higher specific impulse allows a significant reduction of the IMLEO. A detailed comparison between nuclear and chemical propulsions systems is provided in the report describing the 2009 NASA reference mission, which demonstrates the clear advantage of utilizing Nuclear Thermal Propulsion, albeit at the price of an expensive mission [19]. Overall, nuclear-based propulsion systems, thermal (NTP) or electric (NEP), are key technologies for the future of space exploration [14, 25].

In particular, if the mission to Mars is considered as the first of a sequence, the cost to develop NTP or NEP is an investment that will be recovered over the number of follow-on missions. In this sense, it is possible to state that **if humankind wants to become a spacefaring civilization, it must develop nuclear propulsion**.

The main drawback of nuclear propulsion is that its TRL is lower compared to chemical propulsion. However, NTP has been developed in the US and USSR since the 1950s. The tests performed in the 1960s and 1970s on prototype nuclear thermal thrusters demonstrate that the development of a safe, reliable thruster of this type can be performed in a reasonable time and at reasonable cost.

For NEP, the problem of the TRL may be split into two parts, namely the design of the power plant and the design of the thrusters. In the former case, the technology is already fairly advanced and a lightweight nuclear reactor for power generation must be developed for powering the Mars outpost anyway (see below). Electric thrusters are a mature technology in so far as small thrusters are concerned. Research aiming at scaling them up, and improving their performance is underway – in particular the VASIMR has reached a TRL sufficient to plan tests on the ISS. Self-field magneto plasma dynamic thrusters already have achieved significant steps at very high power, for example they have been tested in the range 200 kWe to 1MWe at the Institut für Raumfarhtsysteme in Stuttgart, at Princeton University, the Keldysh Research Center, Fakel, Energya, and Moscow Aviation Institute where a lithium propellant Lorentz force accelerator has been tested at 500 kWe for 500 hours. Provided that this research work proceeds, it appears that at least one of these kinds of thruster will be fully operational before a Mars mission is planned.

Ground testing, however, presents a problem for NTP research as it is unlawful to test the whole system on Earth without a facility capable of capturing any fission product releases. Such an underground facility has been conceptually designed at LASL and has been also declared available in Russia. Additionally, based on the results of previous tests, a test in a safe orbit is neither impossible nor unreasonably costly. Using nuclear propulsion would make it unnecessary to rely on certain technologies that are not yet at a sufficient TRL (like aerocapture), and makes all limitations linked with IMLEO less stringent.

A very important point is that nuclear propulsion may allow faster transit to and from Mars, therefore reducing the crew exposure time to radiation and microgravity. If the problem of radiation during the months spent in space proves to be a insurmountable obstacle and fast transit proves to be required, nuclear propulsion would become the only possible choice.

Nuclear propulsion allows for the design of round trip spacecraft that could, theoretically, be refurbished and re-used for follow-up Mars missions. Multiple journeys could thus be performed in a cost-effective way, provided the nuclear reactor fuel can be replaced. Such a spacecraft would not re-enter the Earth atmosphere from a hyperbolic trajectory, but would enter a LEO (or a higher orbit). The crew can then spend their quarantine either on Earth, after returning immediately from orbit, or in a space station, reducing dangers of back contamination. Moreover, even if chemical propulsion may prove affordable for a flag-and-footprint style mission, more complicated missions involving the building of an outpost, and true colonization, will require nuclear propulsion. If a number of subsequent missions are planned, the cost of developing nuclear (thermal or electric) propulsion can be considered as an investment that is repaid by the lower cost of mounting several individual missions. At the end, whether or not nuclear propulsion is affordable depends on the ultimate purpose of a human Mars program. Finally, once developed nuclear propulsion would support easier exploration, both human and robotic, of the whole Solar System.

The choice between nuclear and chemical propulsion – and if the former is chosen, that between NTP and NEP or their combination – does not depend only on considerations strictly regarding propulsion, IMLEO, the number of missions or spacecraft reusability. Other considerations may play a significant role. Among them, the timeframe for the Mars mission (the longer this timeframe, the more likely that nuclear propulsion will be available if it is recognized to be indispensable), the cost of orbiting payload (which can greatly decrease if more private organizations enter the space business), and the development of space tourism or asteroid mining. These considerations not only enable a more informed choice between chemical and nuclear propulsion, but also between NTP and NEP missions. Developing a plasma thruster to re-boost the space station, together with a nuclear power generator for space station or lunar use, can upgrade current NEP concepts to a higher TRL.

6.6.5. Reasons for choosing chemical propulsion

Choosing the Mars mission propulsion system based only on ISP is unfair. There are many hidden problems and drawbacks with nuclear based propulsion systems. The best criterion on which to make a fair comparison is the IMLEO. Ultimately, whatever the mission, the final choice of propulsion will be political, taking into account total cost, the duration, and costs of the preparatory missions. As for the mass budget, there are several important issues that must be considered:

- It is not possible to send a nuclear engine to LEO due to the risk of reentry (for instance, the ISS is regularly reboosted to its orbit to avoid reentry). Therefore, for safety reasons, any nuclear engine should be parked in a high orbit probably around 1000km, compared to around 400 km for a chemical system. The consequence is a propellant mass penalty for nuclear modules and payloads.

- Radiation from engine operation requires a heavy radiation shield. In the NASA DRA5 report [19], the estimated shield mass is ten tons. A shield is not required with chemical propulsion systems.
- Nuclear thermal rockets use hydrogen as propellant. Storing hydrogen in space for long periods poses many problems. Another problem is the size and mass of the tanks, an important issue when planning aerobraking to capture in a Martian orbit: if the tank is very big, it makes aerocapture more difficult. The heat shields are much heavier, control of the vehicle is much harder and the strategy is not very efficient. The consequence is another important mass penalty. This particular problem also exists for huge chemical propulsion vehicles, but it can be mitigated by the use of several vehicles or by the reduction of the payload that is sent to Mars (e.g. smaller crews). In other words, if aerocapture is enabled by appropriate mission architecture options, propellant requirements are strongly reduced, which makes chemical propulsion systems much more attractive. In the DRA5 report [19], nuclear thermal propulsion is preferred but the comparison is not fair because no effort is made to enable aerocapture (with aerocapture the gain is estimated at 200 tons [24]).
- There exist several mission architectures based on chemical propulsion systems that outperform the NASA reference mission for the IMLEO criterion [22, 34]. This is mainly due to optimizing all parameters of the trade space [24]. The problem with nuclear based architectures is that optimization is more difficult (e.g., feasibility of aerocapture with large hydrogen tanks).
- An important parameter of any nuclear based propulsion system is the specific power that can be achieved. If the nuclear reactor is very heavy, whatever the specific impulse, even if the mass of propellant is negligible, it will not be competitive with architectures based on other propulsion systems for a Mars mission. The problem is therefore not only to build a nuclear based propulsion system, but to build a light and still very powerful (e.g., 10 MW) one. For that reason, even if the TRL of nuclear based propulsion systems is not too low, the TRL of very light nuclear reactors (which are highly desirable for the long term and are game changing technologies for the future) is low.
- SEP architectures are an interesting propulsion alternative. The technology is already mastered for small probes, the maturity is relatively high and the development of 300 kW solar tugs or close to that would probably be fast with affordable investments. However, there are four important drawbacks with SEP:
    - Low thrust escape trajectories result in large gravitation losses reducing the benefits of using high ISP SEP.
    - Aerocapture maneuvers with giant and fragile solar panels is unfeasible. There is therefore a mass penalty for braking, also reducing the benefits of high ISP SEP. Once again, the IMLEO is the only parameter that should be taken into account for fair comparisons.
    - In general, since reaching escape speed with SEP systems takes too much time, it has been suggested in the past to send the deep space habitat to

high orbit without the crew and to board it when convenient using a small rocket-powered capsule (a 'space taxi'). This would require a complex and more expensive organization.

- In order to speed up the outbound trip, it might be necessary to use chemical propulsion systems or a nuclear thermal rocket. SEP might therefore be combined with other propulsion systems.

All in all, nuclear based propulsion systems seem appropriate if large spaceships have to be sent to Mars. Several important options for chemical propulsion have not been investigated and considered in comparisons found in the literature (see next sections).

Concerning the "fast transit" argument in favor of nuclear propulsion, this is biased for two reasons:

- Since the TMI burn is provided in LEO, the ΔV increase for a four to five month trajectory is not that high. The requirements are presented Table 6.3. It should be noted that it is wrong to say that Mars at aphelion would be out of reach with chemical propulsion as the increase is only 0.3 km/s.

	Hohmann for Mars perihelion	Hohmann for Mars aphelion	Free return (2 years period elliptic trajectory)
V_∞ (km/s)	2,30	3,51	5,08
ΔV 400km (km/s)	**3,42**	**3,73**	**4,31**
Trip duration (days)	237	281	157 (Mars aphelion)

Table 6.3. ΔV requirements for a TMI burn operated in LEO at 400km.

- A key idea to minimize the risks of the mission is to choose a free return trajectory [31]. The spacecraft is sent to Mars on a two-year period elliptical trajectory around the sun, so that it encounters the Earth after exactly two years if there is no Mars landing. This trajectory implies a higher ΔV (compared to Hohmann's trajectory) but the impact on propellant requirements is acceptable (such a ΔV is even higher than the ΔV proposed in the NASA DRA5 [19]).

It is claimed that the recurrent cost would be highly reduced using nuclear propulsion because the vehicle could be used several times. First, the development cost is possibly a deterrent (see introduction), not the recurrent one. And second, reusability is very challenging in the space domain. Payload carried to LEO by the Space Shuttle was more expensive than by expendable launchers, and operations cost was exorbitant, taking a significant fraction of the NASA annual budget. Hopefully, reusability will be mastered in the future, but it should not be considered routine. The path to reusability will probably be very long.

At the political level the most important criteria to choose the propulsion system are cost, length, and return on investment of the preparatory missions. It is doubtful that a costly and long project could be financed just for the purpose of a first human Mars mission. As a consequence, it is felt that the use of nuclear based propulsion systems would allow regular human Mars missions, once these technologies have been developed and used for other exploration missions (for example heavy probes to the outer solar system [21]). A sustainable roadmap for such a project does not exist.

Concerning the argument of a possible stop of the Mars exploration program after the first missions (Apollo syndrome),It can be stated that the context is much different as in the Moon race only a few missions to the Moon were planned and then nothing. Provided that the first mission is implemented and successful, the international nature of those missions and also the interest by the private sector will certainly enable a long term focus on the red planet leading eventually to permanent bases and possibly settlements.

6.6.6. Reasons for choosing SEP

Electric propulsion systems are attractive because of their high ISP [4, 13, 20, 25,29]. The power source can be a nuclear power plant or a large set of photovoltaic solar cells. The second option is very interesting for several reasons:

- First, solar panels are very cheap, simple and reliable.
- Second, solar electric propulsion has already been used for interplanetary probes (e.g., Deep Space 1, Dawn, Hayabusa). Its TRL is therefore higher than for nuclear power.
- Third, the photovoltaic technology has been improved since its first use in space [6]. Very light and efficient solar panels can be built, making that technology competitive with nuclear power in terms of specific power.
- The use of high-power SEP makes the architecture significantly less sensitive to mass growth and improves flexibility.
- In the NASA reference mission based on NTP systems, the Orion capsule and its service module are underexploited. The capsule is used only at the very beginning and the last day of the mission, while it is capable of supporting a crew during 21 days and it can operate important orbit transfers. Its use is therefore an important mass penalty for the architecture, while a much smaller capsule would be sufficient. In the SEP scenario, the Orion vehicle can be used to transfer the crew to a high Earth orbit and perform a rendezvous with the main interplanetary vehicle. In that case, its size and mass are appropriate.
- For chemical propulsion, the amount of propellant for TMI highly depends on the position of Mars on its orbit, due to its eccentricity. Since propellant requirements are much less for SEP, architectures based on SEP are less sensitive than chemical ones to the relative position of the two planets.
- At some point in the future, electric propulsion systems will certainly play the role of game changing technologies for deep space transportation. It might be difficult to deploy several square kilometers of solar panels to achieve several MW of power while keeping the overall mass low and the robustness high. It might be easier to achieve the same power with ultralight nuclear power plants. Whatever the power source of the future, it will supply electric power to ion thrusters with a high ISP. A 300 kW SEP tug is therefore an important step in that direction.

All in all, SEP might be a good trade-off between all chemical architectures and nuclear based ones.

6.7. Mars orbit insertion

At the end of the interplanetary travel, a Mars orbit insertion maneuver is required. ΔV to insert a spacecraft in Mars orbit are on the order of 2 to 3 km/s, depending on interplanetary coasting speed and orbit. There are three options for Mars orbit insertion:

- All propulsive: a propulsion system with high ISP is used to slow down the vehicle.
- Aerocapture: the vehicle enters Mars atmosphere for intense atmospheric braking. This requires a heat shield [32].
- Aerobraking: similar to aerocapture, but several passes inside the upper Martian atmosphere complete the braking phase.

The third option is generally eliminated because it is time consuming and does not bring important advantages compared to aerocapture (a heat shield is needed anyway and the orbit must be adjusted after each pass). Aerocapture is a risky maneuver because the vehicle has to enter a thin flight corridor. In addition, if the arriving velocity is high, if the vehicle is big or if the shape is complex, it is difficult to protect it and control its attitude during the atmospheric drag. A parametric risk analysis for aerocapture is shown in Table 6.4. It is worth noting that in the last NASA reference mission, aerocapture is proposed for the two cargos but not for the crewed vehicle precisely because of size and shape of that vehicle.

Parametric risk analysis Risk levels from 1 (low risk) to 5 (high risk)	Compact shape	Elongated shape	More complex shapes (a capsule is docked for instance)
Small size	1	2	3
Big size	2	3	4

Table 6.4. Parametric risk analysis for Mars orbit insertion.

The physics of aerocapture suggests the following recommendations:

- Use small vehicles: atmospheric braking is easier for vehicles with a low ballistic coefficient. The braking phase occurs at higher altitudes with a smaller deceleration peak. Big vehicles generally have a high ballistic coefficient.
- Integrate all modules in a simple shape that facilitates aerocapture (conic shape for instance). Avoid modules attached to nodes located on the side of the vehicle (e.g., Orion attached to the crewed vehicle in the NASA reference mission).

These constraints seem to be manageable. The choice for Mars Orbit insertion is strictly dependent on that for propulsion. If chemical propulsion is chosen (or SEP for departure and chemical in the vicinity of Mars), aerocapture must be chosen for all vehicles and the above recommendations will have to be taken into account.

In addition to that, backup strategies may be considered if aerocapture fails or if an abort is necessary (for instance if there is a problem with attitude control or thermal protection). The minimum backup strategy should be able to abort aerocapture and to proceed to a free return or to proceed to Mars orbit insertion using the propulsion stage of a lander, depending on the availability of other vehicles in orbit or on the surface.

Whatever the case, a reasonable recommendation is that a crewed vehicle should have a backup wet propulsion stage to address the eventuality of a problem before or during aerocapture or aerobraking. This propulsion stage would typically be used to avoid hyperbolic trajectories, which would cause LOC.

In case of NEP a propulsive capture maneuver is assumed. In case of NTP, different strategies exist. It is possible to use a mixed strategy: propulsive braking for crew vehicles and aerocapture for cargo vehicles.

Recommendations

- If chemical propulsion is chosen, assume aerocapture in the early design of the mission and propagate the constraints to the follow-on choices.
- Since aerocapture is easier and more efficient with small vehicles and simple shapes, investigate options where the payload is split into several elements stored in smaller and lighter vehicles.

6.8. Descent vehicles and EDL strategies

Cargo and habitat landing is one of the most complex problems of the Mars mission [3].

Some of the parameters to be considered:

- The size and mass of the payloads to be landed:
  - habitat
  - Mars ascent vehicle
  - ISRU Processing Unit
  - Surface power systems
  - Surface vehicles
- The ballistic coefficient of landing vehicles
- The initial parking orbit
- The altitude of the landing zone
- The characteristics of the guidance system to ensure cargo and habitat land in close proximity

Several options exist for EDL. According to recent studies, rigid deployable heat shields or Hypersonic Inflatable Atmospheric Decelerators (HIAD) are promising systems [3, 5, 20, 28]. If the mass of the payload is of the order of 40 tons, the HIAD approach appears to be the more mass effective [20]. However, a large HIAD may have steering issues during the guided phase of flight. Rigid heat shields or smaller HIAD vehicles provide more control but at the expense of size (fairing problems) or mass of the payload.

Because of the complexity of the physical models, the qualifications of EDL systems and procedures might be very long and terribly expensive with the requirement of performing several tests at full scale in the Martian atmosphere. As pointed out in the introduction, this is a possible showstopper for human missions to Mars. Eventually, a solution might be found but at the expense of the complexity of the systems with high uncertainties on the success of that phase. A major problem with EDL is the variability and complexity of the physical models, which makes the estimation of the probability of failure difficult. For such complex problems, maturity models may provide a better estimation of the risks. The probability of success typically increases with the number of successful trials at full scale in the Martian atmosphere. The risks can therefore be decreased below a desired threshold, but very slowly with important implications on the costs and duration of the tests. An important parameter of the maturity model is the decreasing rate of the probability of failure, which depends on the complexity of the EDL phase. For simple EDL configurations, a few tests at full scale could be sufficient, while for more complex ones, dozens of tests could be necessary. If it is required to send tens of heavy vehicles to Mars to achieve an acceptable risk, the project might become unsustainable and be cancelled. As a consequence, EDL is the domain where the efforts have to be made in priority to determine the most simple technologies and procedures, eventually at the expense of strong constraints on other parts of the mission. Parametric risk modeling could be very important for EDL decision making. It is indeed important to understand how the complexity and risks increase with the mass, shape and entry velocity of the landing vehicle [23, 27]. The problem is illustrated with an estimated result plotted in Figure 6.5. If the probability of failure is too high, as it suggested in that figure, the best strategy, if possible, could be to split the payload into several modules not heavier than 16 tons and to land several small vehicles instead of a big one.

According to several authors, for the same initial payload mass, several small landers might be lighter than a single heavy one [5, 28]. See Figure 6.6. This important issue is correlated with the Probability Of Failure (POF). If small landers are less risky and lighter than a big one, there are important implications for the architecture of the mission. Options in which important elements of the payload are sent to the surface separately have to be seriously considered.

Recommendations

- An early parametric risk analysis is required to determine the POF as a function of payload mass, shape and entry velocity of the landing vehicle.
- In parallel, studies should be carried out to determine the payload mass fraction as a function of the entry mass.
- With POF and payload mass fraction estimated, strategic choices regarding EDL systems (e.g., choosing to land vehicles lighter than a given threshold to reduce the complexity of EDL systems and mitigate the risks) might not be possible without strong constraints on other important parameters of the mission, including the size, mass and number of interplanetary vehicles, the ISRU strategy, the size of the crew and the size and mass of the surface vehicles. It is therefore recommended to set the EDL options as a major parameter of the mission and to examine the possibility of adapting the other parameters. The main criterion for decision making is the global risk of LOC.

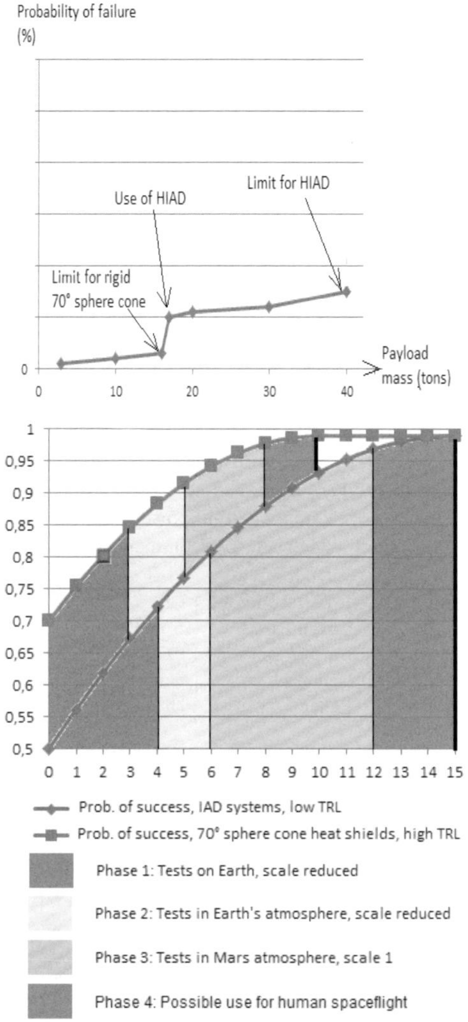

Figure 6.5. Above: Illustration of a parametric risk analysis for EDL systems before the first test at full scale in the Martian atmosphere (estimated plot). Bottom: Maturity models can be used to determine the number of tests that are required to achieve an acceptable probability of failure (this graph is for illustration only; the exact profiles have to be determined by experts).

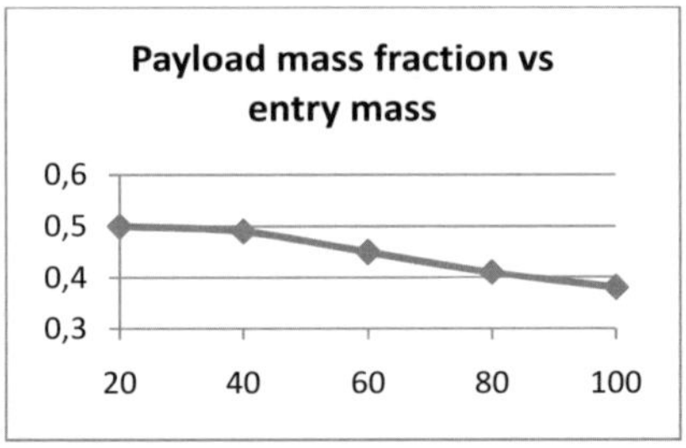

Figure 6.6: Payload mass fraction as a function of the entry mass in tons [20].

6.9. In-Situ Propellant Production (ISPP)

Most studies suggest the use of methane and oxygen as propellants for the Mars ascent vehicle [17, 19, 33]. The main options are presented Table 6.5. Some sizing considerations are provided in the NASA reference mission [19]. The most mass-efficient option is to extract water from the soil (at least 3% expected) and carbon dioxide from the atmosphere to produce methane and oxygen. This requires robotic excavators, well-known chemical reactions and a power source. ISPP allows a significant reduction of landers mass, which is also recommended to minimize risks.

Propellant for Mars ascent: $CH_4 + O_2$	No carry-on resources. H_2O from the ground, CO_2 from atmosphere + chemistry	H_2 ferried from Earth CO_2 from atmosphere + chemistry	CH_4 ferried from Earth CO_2 from atmosphere + chemistry	Both CH_4 and O_2 ferried from Earth
Benefits	Landers mass minimized.	Some mass savings.	O_2 from atmospheric CO_2 is relatively easy and robust. Some mass savings.	Independent from local resources and complex ISRU systems.
Drawbacks	H_2O extracted with complex robotic processes. Automation might be difficult. Human presence probably required.	In transit H_2 losses or complex cryogenics. Cryocoolers required.	Heavy cargo lander.	Very heavy cargo lander and impact on IMLEO.

Table 6.5. Four ISRU/ISPP options for the production of methane and oxygen.

The mass of ISRU systems (power included) is less than 25% of the mass of propellant required for the MAV. A key drawback is the use of robotic excavators that probably requires the supervision of the work by humans. Interestingly, that option was discarded in the NASA report and the presence of humans for the deployment of ISRU systems and the supervision of the work has not been considered [19].

Recommendations

In-depth studies are required to assess feasibility and risks associated with the extracting H_2O and CO_2 to manufacture propellants. The presence of humans is probably necessary to ISRU/ISPP operation.

If H_2O extraction is predicted too difficult or if the MAV must be ready for take-off before the departure of the crewed vehicle, the best trade-off is probably to bring the methane from Earth and to produce oxygen using in situ resources, as suggested in the NASA design reference architecture.

6.10. Mars orbit to Earth orbit strategy

In terms of energy, one of the main challenges is to send a habitat and a wet propulsion system from LEO to Mars orbit in order to prepare the return. In addition, since the re-entry in the Earth's atmosphere requires a potentially heavy heat shield, whose mass is correlated to the size of the vehicle, it is usually assumed that a small re-entry capsule is docked to the main habitat and is used by the astronauts the last day of the mission. The total mass of that Earth return vehicle cannot be less than sixty tons (optimistic estimate). In order to send that payload from LEO to Mars orbit, a heavy chemical propulsion system would be required and the total would largely exceed the payload capability of a single heavy launcher. Different strategies can be followed to avoid the assembly of a huge vehicle in LEO.

The first strategy is to choose a propulsion system with a high ISP in order to reduce the amount of propellant [19]. In general, the same habitat and propulsion systems are also used for the outbound leg of the mission, which results in stronger requirements. The assembly of the vehicle in LEO is not avoided but the different modules generally require no more than two or three launches of a 100 ton payload.

In all chemical architectures the assembly of a huge ERV in LEO is generally avoided or limited thanks to other options. Assuming a 130 tons LEO capability for the launcher (maximum LEO payload of the NASA SLS Block 2), the maximum payload for a direct TMI maneuver with chemical propulsion is 46 tons and it is around 40 tons for a fast interplanetary transit. Crew size and aerocapture allow some mass savings; in order to further reduce mass, three strategies can be followed:

- Bring more materials from the Martian surface and optimize the list of elements that are waiting in Mars orbit. Zubrin proposed a direct return from the Martian surface but the mass of propellant that has to be manufactured on Mars would probably be too large [33]. Salotti proposed to send the main habitat back to Mars orbit and to store the consumables for the return in the small capsule [22]. Another option is to send a wet propulstion system to Mars orbit.
- Split the ERV into 2 modules, launch them separately from LEO and assemble them in Mars orbit. This may not be complex if modules are not too big. It is preferable to perform the assembly in Mars orbit rather than in LEO because it

eliminates the need for a large propulsion system, which would also have to be assembled and maintained in orbit with the help of re-boost modules.

However, these strategies do not have to be implemented at the expense of the mass of the other vehicles, and especially the landers. The optimization of the mission suggests that all modules and materials that are required for the return but are not required on the surface of Mars should stay in Mars orbit.

Recommendations:

If high ISP propulsion is chosen, use the same crewed vehicle for the outbound and inbound legs of the mission.

If chemical propulsion is chosen, make aerocapture a priority and try to optimize what is sent to Mars orbit from LEO to prepare the Earth return vehicle.

Two options are recommended:

- Proceed to the assembly of the return propulsion system and habitat in Mars orbit rather than in LEO,
- Send to Mars orbit :
  - The propulsion system and the propellant for the return (TEI);
  - The small capsule for re-entry in the Earth atmosphere;
  - The consumables for the inbound trip (in the capsule).
- Send the main surface habitat back to Mars orbit and leave it there at the end of the stay.

6.11. Redundancy and multiple missions strategy

Ideally, there should be at least one backup habitat module at every moment of the mission and one viable backup strategy ready before each mission phase. However, providing many backup strategies might add a lot of complexity and cost and could increase risks.

A strategy increasing redundancy consists of splitting the payload into several modules. The first idea is to use very small modules with only two astronauts per habitat and to duplicate the entire mission as proposed by Salotti [22]. This would provide many backup scenarios.

Another idea is to exploit all modules as follows:

- For the stay on the surface of Mars, it is assumed that there will be one cargo module and one habitable module. A backup life support system can be added to the equipment of the cargo to provide redundancy.
- For the outbound trip, the ERV or the cargo can play the role of the backup vehicle.
- For the inbound trip, the small capsule can be the backup habitable module.

Recommendation

Examine backup strategies early in the design of the architecture.

Chapter 7
PLANETARY INFRASTRUCTURE AND VEHICLES

7.1. Basic elements and infrastructure development operations

Human Mars exploration requires a variety of equipment to be carried to the planet. Transporting all the required equipment in a single spacecraft is costly and practically unfeasible, and even less so if astronauts are on board. This is impossible if Mars Orbit is attained through an aerobraking maneuver.

Many scenarios call for splitting cargo missions between at least two launch periods, and astronauts only after the essential pieces of equipment are safely installed and working on Mars. Surface equipment depends on mission purpose and Mars exploration strategy.

The design of the equipment depends on the mission goals also in subtle ways: if several missions are to land in the same site to form a sort of permanent infrastructure, they must be designed for durability and ease of maintenance. If each mission must land in different locations, the equipment must be considered 'disposable'.

The surface elements to be delivered to Mars for a typical human mission are the following:

- Habitats,
- Power systems,
- ISRU plant,
- Rovers,
- Workshops, greenhouses, and auxiliary equipment
- Landing guidance system.

Some of them, like the power systems, are compulsory; others, like the ISRU or the greenhouse, depend on whether they are required for the chosen mission architecture.

7.1. Surface Habitat

To sustain a long term human presence on Mars, the habitat must not only protect from the environment but also be built in such a way to encourage a small society to flourish. Long-stay missions will last about 460 days on Mars surface, but in case of several landings in the same place, a much longer useful life may be required.

The Surface Habitat shall support astronauts living and working conditions through the following functions:

- Providing sufficient volume per crew member
- Providing a shirt sleeve environment and a breathable atmosphere
- Providing protection from the Martian environment
- Providing food and water
- Providing means to perform scientific missions and other tasks

The configuration and technology of the Mars habitat module is driven by the lander architecture. Due to re-entry constraints, the available volume per crew member inside the lander is less than ideal. This suggests the need for an inflatable habitat.

The surface habitat shall also provide means to Martian ExtraVehicular Activities (EVA). They include a traditional airlock (maybe inflatable) or a so-called “Suitlock”: an external suit with a mechanical interface on the back-pack that allows berthing with the primary structure of the habitat. A suitlock reduces contamination hazards and airlock mass.

When designing a Mars base, it is important to take into consideration the following factors.

7.2.1. Physical factors

Air

The air quality inside a habitat or any interior space built on Mars must be comparable to the one on Earth. This takes into account oxygen pressure and humidity, to be regulated by the Environmental Control and Life Support System (ECLSS). The atmospheric pressure on Mars is about 0.006 bar, less then 1% of the Earth's. Keeping the habitat interior at the same Standard Temperature and Pressure (STP) conditions found on Earth places a significant load on the base structure. At ground level Mars temperature ranges between –63°C and +20°C in the southern hemisphere during mid-summer and can drop down to –100°C [1]. Because of the thin atmosphere, thermal insulation is not a problem since conventional materials should be sufficient.

Cosmic Radiation

Cosmic Radiation is the biggest problem related to spaceflight and colonization. Mars has limited protection against it and adequate protection must be ensured, especially for long stays. Materials offering the best protection against radiation are hydrogen rich materials, like water, hydrogen fuel or even some plastics currently being developed at NASA. Martian soil (regolith) can also be used. Perhaps it is also possible to take advantage of caves: large lava tubes have been detected where habitats even of large size may be built with sufficient equipment.

Energy

Energy must be produced not only for heating and ECLSS, but also for communications, computers, scientific instrumentation, etc. These need much less energy than ISRU, ISPP, or charging rover batteries.

Mars has an average solar radiation of 589 W/m^2 (371 cal/cm^2/sol) [2]. It is commonly assumed that solar power is insufficient to a human Mars base. If nuclear power is considered, a reactor is required since radio thermal generators, although currently used in several space missions, have high mass/power ratio and depend on isotopes (like ^{238}Pu) which are extremely expensive and in very short supply. The true alternative is thus between solar panels and a nuclear reactor. Since a single power station for all the functions of the outpost is needed, this will be discussed in a specific section.

Water/Nutrition

Although it is possible (but costly) to send to Mars all water and food needed by the stay, a closed loop life support system including food and water has been suggested. The amount of resources needed – food, water, gases – is affected by mission duration and number of crew. ECLSS technologies may reduce their mass by recycling, recovering for example oxygen from metabolic carbon dioxide and water from urine and habitat moisture. Loop closure can be increased by producing food in greenhouses and recycling solid waste. The proper level of ECLSS closure shall be decided following the choice of the number of crew, mission duration, power consumption, mass budget optimization, and – in the end – feasibility and cost.

7.2.2. Human factors

When on a long mission, such as a Mars mission, the comfort of the crew should equal that on Earth. The psychology of the astronauts is a key issue (see Chapter 5). This will probably require candidates to be screened in greater depth than done for ISS missions.

Public/private spaces

The naval practice of "hot bunking", where several crew members take turns using the same bed, should be avoided.Humans simply need a space of their own. It is extremely important for the crew's psychology that every crewmember has an individual crew quarter for retreat. It is also crucial to have a common space in which meetings can be held but both aspects must be clearly separated. Recreational space is also recommended. There should be a common space where conventional common recreational activities are possible. There should also be some semi-private or semi-public spaces that are not completely isolated, which offer a small one-on-one meeting space; like, for instance, a niche to look out of the window or a space to simply relax and rest for a while.

Summarizing, there must be:

- Private and individual space for everybody
- Public meeting room
- Semi-public / semi-private recreational spaces

Health

As there is no possibility of returning a crewmember to Earth to treat a medical emergency, the outpost must cater for these kinds of emergencies. There must be a small sick bay that is able to deal with both small emergencies and long time care. Small surgery must also be an option in case of a serious problem. Approximately 6.5 m^3 must be kept free for medical equipment in such a facility [3].

Summarizing, there must be

- An adequately equipped medical facility and
- The possibility of dealing with small emergencies, long-time care, and basic surgery.

Society

The structure of the base must reflect or be a support for the structure of the society. The resources must be shared according to a strict distribution system. Every Martian habitant will have to occupy a crucial role in managing the base – looking after the people, the resources, and the vital systems – or as a researcher. The tasks should be varied, having every crewmember contribute in every field.

Program/function

Water and food must come from a closed loop life support system. It is important for the moral of the crew that the food tastes good and ideally has some fresh ingredients, such as herbs or other greens, as well. A fully equipped kitchen including a freezer, oven, a sink, and cooking and eating supplies are required. [3]

Water is required to be used as drinking water but also for the kitchen, for cleaning the habitat. A hygiene facility including a toilet and some hot, wet towels are recommended, as well [3].

Exercise is important to maintain physical health. Equipment such as a treadmill and elastic bands are beneficial to have.

An area for personal or common recreational activities should be planned. Personal material is also needed for recreation. Things such as watching movies, access to news on Earth, e-books and other conventional recreational activities as on Earth should become part of the activities. Still some space will be needed for paper and other hardware. Currently, on the ISS the astronauts' favorite activity is looking at their home planet through the window. Although this will be difficult from Mars, windows should be included to look at the Martian landscape. Additionally, direct transmission screens should serve as a contact to the outside world, these are particularly important if the habitat is located in a cave or lava tube. Further, each hatch to the outside should be equipped with either a window or a transmission to the outside environment. This enhances the safety when crewmembers are on EVAs and close to the habitat.

For surface exploration, there must be an airlock with a suitport and access to a rover for extra vehicular activities.

In summary, equipment such as a freezer, an oven, a sink, a dispenser, cooking and eating supplies, hygiene facilities, exercise equipment, television, e-books, recreational activities, windows, airlocks with suitports and access to pressurized rovers, etc. must be supplied.

The areas or rooms needed are:

- Galley or kitchen,
- Crew quarters,
- Meeting/communal space for social gathering, dinner, recreational activities,
- Laboratory,
- Work spaces (computer and bench with 3D printer),
- Greenhouse,
- Hygiene facility,
- Medical facility,
- Storage spaces,
- Space for the life support system
- Airlocks, and/or hatches.

Some of these areas may be better separated from the main habitat, like the greenhouse, the lab or the workspaces, which may also be the garage for the rovers, and they will be dealt with in specific sections.

<u>Automation</u>

A functional Mars outpost or habitat should be fully automated. Humans only have a certain amount of force and attention span, especially over a long period in partial gravity [3]. Therefore all systems must be automated and considered at least complementary to human capabilities or, ideally, capable of replacing them.

Contacting Earth can take fourty-four minutes but becomes impossible during the few weeks when the Sun is between the two planets (solar conjunction). Mars Curiosity's rover, for example, is controlled from Earth when possible but has been designed to be autonomous enough to drive on unknown terrain to the next destination. The Mars habitat should be designed with the same principles in mind.

Automation also plays a crucial role when setting up the habitat. A possibility is to send the Mars base and astronauts together in the same mission. Astronauts would do EVAs in Mars orbit to separate the base from the interplanetary habitat. This is a dangerous task that, to be done in a short amount of time, needs a significant amount of automation. The second possibility is to launch the habitat one and a half years before the astronauts. Although this solution is safer, it requires a fully automated and teleoperated procedure for assembly.

Design must consider:

- Automation during the set up of the habitat on Mars,
- Human shortcomings in 1/3 g,
- Automation to compensate for the communication delay between Earth and Mars.

<u>Multifunctional spaces</u>

Flexible spaces and spaces that can be repurposed for multiple uses are a necessity because of the need to allow the crewmembers to personalize their space. Crewmembers might want to adapt and add simple changes to the habitat. This is valid for both private and common spaces. Spaces that do not serve a predefined purpose can be used to create more or less open areas. This allows the visual spatial impression to change which can also a change 'atmosphere' of a room. It is advisable that the lighting, air-conditioning and even colors and textures will be controllable and changeable. Sliding or foldable walls might also be a solution for allowing the architecture of non-fixed spaces to change.

It is advisable to grant:

- Flexibility that allows the crew to modify the space,
- Flexibility that allows the crew to personalize the space,
- Spaces that do not serve a predefined purpose.

Human-Machine Interface (HMI)

The equipment should be designed in a way that it is intuitive and efficient. Machines that are connected or share a common task should be positioned in a way to minimize transitional paths. Critical controls must be accessible from more then one location and there should be an easy access for maintenance and repairs.

It is possible to use portable computers to access and control systems. This has the advantage of eliminating some of the dedicated workstations, displays and controls. However, remote controlled activities and other activities with special requirements will need workstations that include the necessary controls, materials, and tools. This applies, for instance, to remote controlling a rover or a laboratory.

For EVAs, a rover containing two persons should be included as a part of the base's infrastructure. Its assembly can rely more on human work, as it is not critical to survive, but there must also be a room to control these activities and even command the rover from a distance.

Important points:

- Minimizing transitional paths: machines that are connected, or serve a common task should be close to each other,
- Access to main controls from different locations,
- Easy access for maintenance and repair,
- Portable computers minimize the number of workstations,
- Rover and workstation for controlling the rover.

Habitat Modularity

As the number of people and materials that can be brought to Mars is limited, it is useful to plan a modular base to be expanded by adding modules.

Psychosocial aspects

A crewmember will typically have to be available for work about ten hours per day. There must be a rest day every seven to ten days. Fourteen hours must be kept free for sleeping, eating, personal hygiene, recreation, exercise and private communications. A crew is the most efficient if the work is varied. This means that errors and accidents are limited if, for instance, one day the activity is physical and the following the activity is mental. Extravehicular activities should not be allowed two days in a row because of fatigue [3].

Physical and psychological limitations result from being confined to a small area. The crew must be able to have open spaces and individual spaces. Over time, people tend to prefer more complex and larger spaces. Adding windows and integrating loop-like designs can give an impression of larger spaces [4]. Spaces also appear larger when partitioning is kept to a minimum. The windows give the impression of watching, rather than being watched which decreases the feeling of confinement by extending the visual horizon. Curved walls also enhance the impression of a space being perceived larger than the actual size because the borderlines of horizontal planes and vertical walls dissolve.

Humans tend naturally to install their personal space in regular patterns, with the most space possible between the private spaces of different persons. On Mir space station, astronauts usually abandoned their private quarters being right next to each other and installed themselves in other spaces. This is applicable for both sleep and private workstations. Not only must the private space be isolated in terms of sound, vibrations, light, and view, they must also isolated spatially. Physical proximity of the crew quarters should be avoided.

Considerations worth to be taken into account are:

- Work hours are limited by different factors
- Architecture should minimize the impression of confinement
- Physical proximity between the different private quarters and the workstations should be avoided and
- Private quarters should be isolated in terms of sound, vibrations, light and view.

Safety

A workshop in which it is possible to repair damaged materials is required. All tools for maintenance and repair must be found in the habitat and there should be a redundancy for critical material. There must be at least one 3D printer on the base as it can print basic tools, which reduces the required redundancy and the risk in case of failure of basic tools.

When designing an outpost on Mars, it is important to keep in mind the different pathways used. Activities occurring simultaneously should be far apart enough not to interfere with each other or to have crossing pathways that could pose a hazard. Also, emergency exits must be as direct as possible from anywhere in the base. All systems should be designed following fail safe criteria and be redundant.

Safety considerations suggest:

- Presence of a fully equipped workshop for maintenance and repair
- Redundancy for critical and irreplaceable material
- Presence of a 3D printer
- Minimizing crossing pathways
- Emergency exits accessible from all locations, remembering that exit is possible only wearing space suits.

Usability

It is extremely important to have a base that does not transmit noise and vibrations. If it is not possible to obtain silence in some areas like the sleep area, the effect on the crew can vary from anxiety and poor performance to hearing impairment. The toilets and the exercise area should be far away from the beds, as it is the loudest equipment.

The standards now demand for all equipment, like seats and tables, to be adaptable to people ranging from 150 cm to 190 cm in height [5]. For an optimal comfort, these should be adjustable.

For a mission longer than six months, a habitable volume per crewmember of about 25 m^3 is very strongly advised[6]. Crowding and a lack of free volume is a significant factor in crew error [3]. In partial gravity, like on Moon or on Mars, the entire volume is no longer equally accessible, as it is in space, and horizontal surface area becomes more important. Also, some area is lost because beds and surfaces must be horizontal. This must be factored in when calculating the volume and area needed per crewmember.

There must be:

- Good isolation in terms of sound and vibrations,
- Furniture adjustable to individual crewmembers,
- A minimum of 25 m^3 of accessible space per crewmember.

Transportation constraints

Because of cost, mass and volume of Mars habitats should be minimized. Volume is constrained by the launcher payload fairing. As such the components of the outpost must fit into an estimated 6 m diameter payload shroud. The habitat may be deployable, inflatable, or assembled from several parts on Mars, ideally in an automated, self-deployable way.

7.3. Power Plant

Apart from the life support system, much energy is required for ISRU, rovers, scientific equipment, and all other devices which may include lighting a greenhouse.

A comparison of the different surface power strategies has been recently carried out and is available in the literature [5]. There are essentially two alternatives: a solar or a nuclear plant. A solar power system must consist of non-tracking, thin-film roll-out arrays and either batteries or regenerative fuel cells for energy storage. A fission reactor is the second alternative.

The reasons which compel to use very large solar arrays are:

- The distance of Mars from the Sun,
- The difficulty of keeping the solar panels oriented sunwards – probably thin-film arrays should be laid on the ground or on the habitat top,
- The need to have 24-hours energy available, which should be considered with the required efficiency of the associated energy storage devices
- The necessity to provide energy during dust storms, and the deterioration of the solar panels with time.

Additional problems are due to dust removal after a dust storm either robotically or by the astronauts.

A solar power system seems limited to outpost configurations where the energy requirements are limited, while nuclear power ensures continuous power if ISRU, ISPP, greenhouse and equipment including digging and robotic rovers are to be operated. This is particularly true if the surface crew is larger than three three. A nuclear powerplant is also more tolerant of dust than solar arrays.

In particular, the above mentioned equipment and the charging of the batteries of the vehicles – if they are electric – operate also in the night period, when a nuclear power plant is still supplying power, while photovoltaic cells are not working (and should therefore supply much more power during the day to compensate for the difference).

Current designs of surface nuclear power generation are monolithic, both because of power output (in the 0.01 to 1 MW range) and transportability. A monolithic nuclear power system shall include:

- The reactor,
- The thermal-electrical conversion system,
- The cold source (radiator),
- The power conditioning and distribution systems and
- The power storage (battery and/or fuel for fuel cells)

The system is best buried to ensure radiological protection and to save shielding mass. Conventional fuel and architectures limit development cost and risks. The power conditioning is driven by:

- The distance from power plant to habitat. Radiological practice indicates 100 m as a safe distance if the reactor is buried in regolith, a distance which does not make any problem, but may rule out low voltage DC transmission and
- The choice of frequency, AC or DC operation or combination
- Output voltage. This depends on conversion technology and application. Typical aircraft voltage is 28V but may be higher to limit transmission (ohmic) losses. On the ISS, for instance, the secondary power distribution is at 124.5 Vdc or, in the Russian segment, at 48 Vdc.

It must be noted that safety concerns for surface nuclear power are much less politically justified than, say, operating a reactor in Earth orbit, since there is no possibility that the reactor returns to Earth after becoming operational.

7.4. ISRU plant

In-situ resources may become the crucial factor to sustain human missions on Mars. Chapter 6 listed four ISRU strategies (see also Table in Appendix B):

- Bringing CH_4 and O_2 to the surface of Mars;
- Bringing H_2 and exploit Martian atmospheric CO_2 to produce CH_4 and O_2;
- Bringing CH_4 and exploit Martian atmospheric CO_2 to produce O_2;
- Exploiting Martian atmospheric CO_2 and Martian water ice to produce CH_4 and O_2.

Trade-offs are complex, impacting on mission mass budgets. There is still no consensus on the preferred option.

ISRU equipment will be landed on Mars before crew arrival. It will process atmospheric CO_2 to extract O_2 for astronauts and oxidizer for the return vehicle. The technology will use Reverse Water Gas Shift (RWGS) to produce O_2 and the Sabatier cycle to produce CH_4.

The main components of the ISRU plant are:

- A RWGS reactor which reduces CO_2 by means of hydrogen;

- A water separator to separate water from CO
- An electrolyser able to split water and produce O_2 and H_2

ISRU development should start with an automated test bed demo; a pilot propellant plant will follow. After testing, an ISRU plant will eventually be transported to complement resources ferried to the surface. Fuel cells and batteries add redundancy to this strategy.

Oxygen and methane extracted from the Martian atmosphere can eventually power the MAV for the crew return to Earth. The main components of the ISPP plant are:

- A Sabatier reactor which reduces the CO_2 by means of hydrogen;
- A separator to remove the water from Methane;
- An electrolyser able to separate Oxygen and Hydrogen (fed to the Reactor) from water;

7.5. Rovers

Rovers are essential to extend areas of human exploration from the landing site. The outpost needs one or more of the following:

- A small robotic or teleoperated rover,
- A small human-rated, unpressurized roving vehicle,
- A pressurized roving vehicle,
- A large pressurized mobile habitat.

One or more small robotic rovers could accompany astronauts during EVAs, scout ahead of human-carrying rovers, or enter areas off-limits or too dangerous to humans. The degree of autonomy of these rovers may vary from fully autonomous to simply teleoperated. Robotic swarms can also be made autonomous. Owing to the proximity of humans, the fully teleoperated option is feasible and enables faster travel, the main limitation here being that the most precious resource on Mars will be astronaut's time and it is questionable whether it should be spent in driving rovers. Increasing the autonomy of rovers remains thus an important goal.

In particular, if the landing site will be the same for all missions, it is conceivable (although costly) to land robotic rovers at interesting points on the planet and to operate them from the outpost. This requires Mars orbiting telecommunication satellites.

Similar to robotic rovers are astronaut 'assistants' accompanying astronauts during EVAs and supporting their tasks. Robots could be built to perform both functions. Specialized rovers may assemble and repair machinery, assist astronauts in their assembly and repairs, in particular of the nuclear reactor. Again the degree of autonomy required may span from fully autonomous robots to teleoperation.

Small unpressurized rovers, similar to the Lunar Roving Vehicle (LRV) of the *Apollo* missions, may carry astronauts during EVAs. The Apollo LRV had a mass of only 210 kg and carried a 490 kg payload. In the last *Apollo* mission it travelled a total of almost 36 km, reaching a maximum distance of 7.6 km from the landing site.

Tests have been performed in Mars analogue sites using quads: small single-seat vehicles similar to quads may be the best choice for short range mobility because of their simple construction, ruggedness and mobility.

A small rover may be carried by a pressurized rover, to allow astronauts access to difficult places. It would be a 'lifeboat' in case of failure or accident to the pressurized rover.

An astronaut 'assistant' could be combined with an unpressurized rover to explore dangerous places and carry astronauts during longer walks or in emergencies.

A pressurized rover is regarded as a key element to sustain exploration by allowing astronauts to traverse larger areas in a shirt-sleeve environment. It should have auxiliary systems, like EVA and robotic tools required to perform exploration tasks in an efficient manner. It could be used in the preparatory robotic phase to support infrastructure development prior to crew arrival.

The pressurized rover is a very complex system because it must meet requirements similar to those for the Martian Habitat, but with long-range mobility. Its mobility may include:

- Crossing the Martian landscape at a nominal speed of 15 km/h, with the capability to reach higher speed where the ground allows
- Climbing 20-degree slopes and their crests (if necessary at a reduced speed)
- Avoiding or negotiating obstacles with a height of 50 cm
- Providing autonomous or remote control and handling

A pressurized rover habitability requirements may include:

- Providing a pressurized shirt-sleeve environment;
- Storing all crew resources, accommodation and facilities for 10 to 14 days
- Accommodating scientific instruments

A pressurized rover may be the size of a van with mass one to two tonsas large as a city bus with a mass of eight tons or more.

A very large pressurized rover would morph, in fact, into a mobile habitat. Enlarging the exploration radius by moving the habitat, even if slowly, is an interesting option. It becomes increasingly difficult with increasing complexity of the outpost. If the outpost is provided with ISRU systems, a greenhouse, etc. with the related power plant, the difficulties related to a mobile outpost become rapidly overwhelming. However, a large pressurized rover may be used as a backup habitat.

The problem concerning the allowable exploration range for rovers is an open one. One possibility is to limit the range of the rover to a distance from which the astronauts can walk back in case of a rover failure. This distance would be severely limited by wearing spacesuits. Two rovers, each capable of hosting the crews of both rovers in an emergency, can increase the exploration range. A light, unpressurized rover is an interesting 'lifeboat' alternative.

Rovers may be powered by rechargeable batteries or by hydrogen-oxygen fuel cells, using hydrogen produced by an ISRU plant. Robotic rovers designed to operate far from the outpost and from astronauts may be powered by radio thermal generators, like the Mars Curiosity rover.

An alternative to electric power is conventional internal combustion engine fueled by methane and oxygen produced by ISRU. This solution has the advantages of high energy density, and relies on centuries old experience with a very reliable and low cost internal combustion engines.

7.6. Workshops, greenhouses and auxiliary equipment

Depending on the length of stay, astronauts must be able to repair vital devices (ECLSS, ISRU plant, rovers, etc.). They will need a purposely designed pressurized zone of the habitat where faulty equipment can be introduced.

Additive Manufacturing (AM) may obviate or reduce the need for spare parts and repairs. Equipment should be designed with this in mind, even if manufactured by different suppliers. A single AM machine should be able to produce all spare parts for a variety of machines.

Design must be coordinated to save mass, starting from materials locally available or at least by recycling waste material. Design updates could be sent from Earth to manufacture improved components.

A greenhouse to conduct experiments and grow food in would improve the crew diet, with the consequent benefits on their physiological and psychological health, while also simplifying logistical problems related to carrying supplies. Technologies to achieve food independence is one of the essential milestones towards reaching the goal of a permanent Mars base. Future terraforming projects would also need these types of devices.

The generic heading of ‘auxiliary equipment’ includes all those devices required by specific aspects of the mission yet not listed in previous headings: digging and construction equipment, fuel tanks and navigation beacons, telecommunications, and many others. These can be specified only during each specific mission planning.

Clearly, most material dealt with in the present section is better suited to subsequent missions than to the first, even though provisions for performing maintenance must always be included. If the overall exploration strategy is based on missions to different locations the mentioned equipment will be reduced to a minimum. If, however, missions are planned to land in the same location and the equipment carried will accumulate on Mars, forming a growing outpost, a large variety of types and functions can be considered. For instance, the first mission can bring just a small workshop; the second a greenhouse and provisions to enlarge the workshop, the third machinery to build roads at least in the most dangerous stretches leading to the locations of interest, etc.

7.7. Space Suits

Different designs and technologies are under evaluation to design a Mars spacesuit. NASA has been working at the so-called Z-series, its next-generation orbital spacesuit. Each iteration of the Z-series will advance technologies for astronauts who must work on the Mars surface. The newest NASA Z-2 prototype has a hard composite upper torso. This composite torso provides the long-term durability required for sustained EVA. The shoulder and hip joints differ significantly based on extensive evaluation performed during the last two years with the Z-1 to look at different ways of optimizing mobility of these complex joints. Lastly, the boots are much closer in nature to those that would be found on a suit ready for space, and the materials used on the Z-2 are compatible with a full-vacuum. Mars space suits will be thoroughly tested to evaluate mobility, comfort and performance in a very harsh thermal and radiation environment.

However, since Mars is not as forbidding as space, suits for Mars EVA should leave more freedom of movement and be more comfortable than suits designed for space EVA. This may imply different space suits for different phases of the mission.

The possibility of designing motorized space suits is interesting and may help reducing EVA fatigue.

7.8. Planetary protection

Both forward (protection of Mars environment from contamination from Earth) and backward (protection of astronauts and of Earth environment from possible contamination from Mars) protection must be implemented. The IAA has started a Cosmic Study on this matter (SG 3.20, Expanding Options for Implementing Planetary Protection during Human Space Exploration).

Planetary protection must be one of the basic concerns when designing all elements to be carried on the Mars surface. In particular biological materials, including human, must be kept separated from the planetary environment.

Forward planetary protection becomes increasingly difficult with increasing variety of biological material on the planet: the presence of a greenhouse, for instance, makes this issue very sensitive.

To prevent backward contamination, from possible biological Martian material, and from the Martian dust and fines, materials exposed to the Martian environment must be cleaned before being introduced into the habitat or pressurized rovers. The dust can be controlled using electrostatic precipitators being currently investigated for future lunar missions, and suit-locks are to be preferred to standard airlocks, where space suits are introduced.

Over the past fifty years or so, significant work has been done towards gaining knowledge about potential biological contamination of solar system bodies, in particular with regard to Mars.

NASA's Planetary Protection policy calls for the imposition of controls on contamination for certain combinations of mission type and target body. There are five categories for target body/mission type combinations but only the relevant Categories IV and V are described here.

Category IV includes certain types of missions (typically an entry probe, lander or rover) to a target body of chemical evolution or origin-of-life interest, or for which scientific opinion holds that the mission would present a significant chance of contamination which could jeopardize future biological exploration. Requirements include:

- Detailed documentation
- Bioassays to enumerate the burden;
- A probability of contamination analysis
- An inventory of the bulk constituent organics
- An increased number of implementing procedures

The latter may include:

- Trajectory biasing;
- The use of clean rooms (Class 100,000 or better) during spacecraft assembly and testing
- Bioload reduction
- Possible partial sterilization of the hardware having direct contact with the target body and a bio-shield for that hardware
- In rare cases, a complete sterilization of the entire spacecraft.

Subdivisions of Category IV (designated IVa, IVb, and IVc) address lander and rover missions to Mars (with or without life detection experiments), and missions landing or accessing regions on Mars which are of particularly high biological interest.

Category IVc. Definies a "Special Region" (from COSPAR 2002, 2005 and NASA 2005) as "A region within which terrestrial organisms are likely to propagate or a region which is interpreted to have a high potential for the existence of extant Martian life forms". In current understanding, this applies to regions where liquid water is present of may occur.

In this context, Committee (SR-SAG) of the Mars Exploration Program Analysis Group (MEPAG) was tasked with investigating the limits to microbial life and the potential for biologically available liquid water on Mars. This committee first specified that, in order to proceed with identifying Special Regions, the definition of relevant words needed clarification. In this context "propagate" was taken to mean reproduction (not just growth or dispersal), while "likely" was taken to imply the probability of specific geological conditions during a certain time period (not the probability of growth of terrestrial organisms).

The results of this study indicated that the definition of a Special Region was determined by a lower temperature limit for propagation (−20°C including margin) and a lower limit for water activity (with margin, an activity threshold of 0.5). Further, a number of remotely sensed features on Mars were categorized as 'Uncertain' (Beaty et al., 2006). Thereafter, a COSPAR Colloquium (Kimneck et al. 2010) recommended that Special Regions be defined by a somewhat lower temperature limit for propagation (−25°C including margin) while retaining an identical water activity threshold of 0.5.

More recently, in the light of a new body of information drawn from multiple disciplines, a further analysis of Mars Special Regions was carried out by a second MEPAG/SR-SAG2 consortium (Rummel et al., 2014). This study included, among other items, a review and reconsideration of the parameters used to define Special Regions and it also provided updated maps and descriptions of those Martian environments recommended for treatment either as "Uncertain" or "Special" regions. These environments include natural features, as well as others potentially formed due to the influence of future landed spacecraft. In the latter context, the committee considered the impact of Special Regions on potential future human missions to Mars, having regard both to the locations of potential local resources and to places that should not be inadvertently contaminated by human activity. Overall, significant advances in the prevailing level of knowledge concerning terrestrial organisms, as well as the capability now achieved to identify possibly habitable Martian environments, have led to a new perception as to where Mars Special Regions may be located and how they should be provided with protection from contamination.

Category V pertains to all missions for which the spacecraft, or a spacecraft component, returns to Earth. The concern regarding these missions lies is the protection of the Earth from back contamination resulting from the return of extra-terrestrial samples (usually soil and rocks). A subcategory called "Unrestricted Earth Return" is defined for solar system bodies deemed by scientific opinion to have no indigenous life forms. Missions in this subcategory have requirements on the outbound (Earth to target body) phase.

For all other Category V missions, in a subcategory named "Restricted Earth Return", the highest degree of concern is articulated by requiring: the absolute prohibition of destructive impact upon return; the need for containment throughout the return phase of all returning hardware which directly contacted the target body or unsterilized material from that body; and the need for containment of any unsterilized samples collected and returned to Earth. Post-mission, there is a need to conduct timely analyses of the returned unsterilized samples under strict containment and using the most sensitive techniques. If any sign of the existence of a non-terrestrial replicating organism is found, the returned sample must remain contained unless treated by an effective sterilization procedure. These Category V concerns are reflected in requirements that encompass those of Category IV with added continuous monitoring of mission activities, studies, and research in sterilization procedures and containment techniques.

Most recently, NASA appointed a dedicated panel to consider organic contamination in the context of their proposed Mars 2020 rover mission. This rover should look for signs of past life, collect samples for possible future return to Earth, and demonstrate technology for future human exploration of Mars. The science conducted by the rover's instruments is anticipated to provide the context needed to make informed decisions about whether to return samples to Earth. A report produced by this panel is contained in Summons et al., (2014) and planning considerations related to the possible organic contamination of Martian samples and requirements for the provision of witness plates, archive facilities and blanks/standards are currently under consideration at a national level.

Chapter 8
THE GROUND SEGMENT AND THE TESTING FACILITIES ON EARTH

The ground segment is particularly important and politically critical in complex and large international missions. Each country will request ground facilities on its own territory, and even non-spacefaring nations may already have some. Ground facilities create jobs and local authorities will try to host them.

8.1. Launch assets

IMLEOs of order of several hundred tons, will require heavy launchers, both for direct launch and in-orbit assembly. The NASA SLS vehicle can contribute to a human Mars mission. The most powerful configuration (Block 2) is expected to have a capability of 130 metric tons in LEO and 45 tons for Mars.

In the same class, existing or in the planning stages, are launchers by China (Long March 9), and Russia (Energiya). Space-X is developing the Falcon Heavy launcher, smaller but nevertheless specifically intended for a Mars mission.

Appropriate vehicle assembly buildings, crawlers, and launch pads will be designed and built by these countries. The question of how to apportion the many launches necessary to an international Mars mission is political and cannot be solved here. This is also the case of key subcontracting work. Maintaining a steady industrial workflow and workforce will be critical, as missions schedule depends on the time constraints for the launch periods, and long time gaps may occur. If there is no rocket to build during a long period, the companies might have difficulties keeping their personnel and their knowledge. Additionally, in most Mars mission architectures, there are important time constraints for the launch periods. It could be required to launch several rockets in the same year. Because it usually takes several months to build and assemble a rocket of that size and the working teams are busy during long periods of time, duplicating the facilities and training new teams might be an issue with possibly important impacts on the complexity and costs of the mission.

8.2. Design and test of key Mars mission objects

In general, a Mars mission architecture requires:

- Interplanetary propulsion systems. Whatever the choice (chemical, nuclear thermal, solar electric, nuclear electric), new propulsion systems will be built and ground tested before space testing. Facilities to test chemical engines and electric propulsion are already available, but for specific needs, it might be required to make important adaptations or to build new facilities.
- A deep space habitat module to support the crew during the Earth to Mars transit and back. The NASA Orion spacecraft can host astronauts for about three weeks at most,

so such module does not exist yet. A specific facility might have to be built for the construction, integration and tests of that module. (See also the section on testing life support systems.)

- Cargo module and MAV. In most Mars mission architectures, a MAV containing an ISRU system will land on Mars before the crew. A specific interplanetary cargo vehicle will have to be designed, built, and integrated in facilities to be determined to accomplish this. The MAV, which is the most important part of the cargo, will be ground tested like the lunar module of the Apollo program was. Martian gravity is only one third of Earth gravity, but the thrust and control of the engines can be tested apart and the MAV mass can be reduced should an ascent flight test be deemed necessary. Orbit testing can also be carried out in LEO.
- EDL systems. The TRL of EDL systems for planetary entry of heavy vehicles is very low. Wind tunnel testing is feasible in some facilities as already done for Martian probes. For human missions the issue lies in the unique mass and size of these vehicles. Scaling using similarity theory may help, but final qualification will require testing in the Martian atmosphere.
- A surface habitat, that might or not coincide with the deep space module. Both should be designed with many commonalities to minimize cost and to reduce the crew's need for cognitive adaptation in the transition phase. Several simple surface habitats have already been built on Earth. However, the habitat module will be optimized to minimize mass, to sustain the crew, and cope with specific environmental constraints: It should withstand low pressure, high temperature excursions, and dust. These constraints might be reproduced in a dedicated facility to be built.
- ISRU and power systems. In Situ Resource Utilization systems will probably be sent to Mars to produce propellants for the MAV or for the production of water and oxygen. Facilities will have to be built on Earth to test and qualify ISRU systems. Surface power systems also need to be designed and similarly qualified.
- Spacesuits. Specific spacesuits for the Mars mission may not need to be designed, like those for astronauts that must work on Mars surface. Existing spacesuits will be tested to make sure that they are appropriate for the various uses during the mission. The impact of dust might be an issue: the Apollo astronauts found moon dust to be an unexpected challenge. It was abrasive, it infiltrated the outer gloves and stuck to everything. The facility testing the surface habitat (low pressure, presence of abrasive dust) might also be used to test the spacesuits.
- Surface vehicles and tools. Whatever the final choice for surface vehicles (unpressurized, pressurized, small or big), these vehicles will have to be tested in similar environments on Earth, and candidate astronauts will be trained to use them efficiently and safely. The same constraints and recommendations apply to tools and scientific equipment.

8.3. Communication centers

Deep space communications are very different from Earth or LEO communications. High gain antennas are required to increase S/N ratio. A NASA deep space network already exists to track planetary probes. It is a worldwide network of large antennas and communication facilities. There are three main sites:

- The Goldstone Deep Space Communications complex in California
- The Madrid Deep Space Communication complex in Spain
- The Canberra Deep Space Communication complex in Australia

The three sites are located around the globe such that at least one of them can orient antennas toward any direction of space at any time. All communication centers are linked to the Jet Propulsion Laboratory control center in Pasadena. They provide the two-way communications link that guides and controls the interplanetary vehicles. The main functions of any Deep Space Network (DSN) are:

- Telemetry and tracking
- Control and command of the space vehicles and satellites involved in the mission
- Communication with the crew.

Other countries have their own DSNs. For instance, ESTRACK is the European Space Tracking network and Russia uses the Soviet Deep Space Network.

As the human Mars mission is meant to be international, several communication centers from different countries could participate to the DSN, providing multiple redundancies.

8.4. Mission control

Specific buildings have to be built for the different teams, which will follow and control the mission. Though the exact organization of those teams have not yet been determined, a long list of expertise is required for:

- Astronautics and flight control
- Power systems
- Mechanics, electronics
- Programming of embedded systems
- Life support systems
- Communications
- Astronauts support (physiology and psychology)
- Science and exploration
- Data storage and processing
- Mission planning
- Command

For every team, all the equipment and facilities that are required for the control and eventual testing and simulations of specific procedures should be developed before they are carried out by the astronauts. The different teams should also be trained to work together.

8.5. In situ propellant production simulations

In most mission architectures, the MAV propellant is produced using local resources, such as the atmosphere to extract carbon dioxide and oxygen [16]. Local resources in the atmosphere or in the soil can also be used to produce oxygen or water [12]. Specific tools (robots, compressors, heaters, coolers) and chemical reactions will be used (Sabatier cycle, water electrolysis, etc.) to accomplish this. Before any test on the surface of Mars, there should be ground testing to determine the best experimental conditions, production rates, storage, power requirements, and system robustness. The chemical reactions are well known and some tests have already been made but not at the appropriate scale [16]. In addition to that, the systems might have to be deployed by robots in an automatically depending on the strategy chosen. Experiments are necessary to test the robustness of the deployment procedures.

Though the Martian gravity cannot be simulated on Earth, the Martian atmosphere and the soil can be artificially reproduced in a building to test full scale ISRU systems.

As for power systems, the automatic deployment of a nuclear reactor or/and solar panels will be tested on Earth. The usability of solar panels after being subject to dust storms are issues for which solutions have been proposed but not yet tested in the field. [11].

8.6. Life support systems

Life support systems for human spaceflight have been developed since the beginning of the space era [9, 15]. However, no mission has been conducted in space without resupply and crew changes over three years. Life support systems on past and current space stations are therefore not entirely appropriate and not optimized for a mission to Mars [6]. Bioregenerative and closed loop life support systems with optimized recycling rates are key issues. The preliminary tests will be performed on Earth. There are also medical issues such as preventing physiological weakness due to microgravity, mitigating radiation effects, and using health monitoring systems [2]. These issues are still being investigated and solutions are still unavailable.

The NASA Johnson Space Center (JSC) has expertise in life support systems technologies (air revitalization, water recovery, waste processing, etc.). It is also specialized in the design and testing of space suits. While most life support systems are based on chemistry and physics, the objective of the ESA Melissa project is to use microorganisms to purify water and design a closed-loop environmentally controled life support system [7]. Several plants based on that concept have been built and used all around the world. This original approach is difficult because living organisms depend on many parameters that cannot easily be controlled. However, the authors claim that some resilience can be obtained. An extensive test campaign will have to be carried out to assess the robustness of such systems.

ENVIHAB is a DLR laboratory specialized in medicine, space physiology and psychology to prepare for future human spaceflight. It is located in Cologne, Germany.

Russia and China have their own life support technologies, and all spacefaring countries can cooperate to the Mars mission from this viewpoint.

8.7. Simulations

Simulating the Martian environment is critical to test:

- The efficiency of systems (life support, recycling, power, space suits, rovers, robots, scientific tools, etc.) and their robustness in a harsh environment and by people with limited mobility (pressurized suits, rigid gloves and others).
- The procedures for surface deployment, EVA preparation, exploring distant areas, using tools, communicating with the habitat and ground mission control, avoiding contamination, removing dust, maintenance operations, rescuing, preparing Mars ascent, and others.
- Human factors: physiological issues (nutrition, health monitoring, medical emergency), psychological issues (confinement, isolation, stress, facing dangerous situations), social issues (collaboration, conflicts, task sharing, leadership) and cognitive issues (selection of crew based on background skills, training, skill development, maintaining competencies). (See Chapter 5).

A number of Mars analog simulations have already been carried out on Earth. A non-exhaustive list is:

- The pioneering Biosphere experiments held in Texas about 30 years ago (Biosphere II). The project partially failed for financial reasons but demonstrated that it was possible for a time to live in a self-sufficient small Earth-like ecosystem
- The Mars Society cylindrical habitats located in different desert areas during several years to test rovers, exploration procedures, human/robots cooperation and to evaluate human factors, (Figure 8.1). Many reports have been published [8].

Figure 8.1. Mars Desert Research Station, Utah, USA (courtesy Mars Society).

- NASA Desert RATS (Research and Technology Studies) has gathered in different desert locations engineers, scientists from NASA centers and partner organizations to hold dress rehearsals for future missions to other planets. Dedicated rovers and robots have been field tested [4, 10]. See also [10].
- Human Exploration Research Analog (HERA). This habitat is an artificial analogue of a planetary base located at NASA Johnson Space Center, Houston, Texas.

- Ground Experimental Complex (NEK). It is an Artificial Analogue of an interplanetary vehicle located in Moscow, Russia. It has been used to study human factors during a 500 days simulation in confined environment [13].
- Hawaii Space Exploration Analog and Simulation (Hi-seas). Originally started to simulate food and culinary routines during long duration spaceflight, but it is intended to be used for more general purposes in the next years [1].
- Pools and undersea assets. Several swimming pools around the world are dedicated to astronauts' training for microgravity. Underwater experiments have been conducted in the Marseille bay by COMEX to simulate EVA activities [14].
- Self-Deployable Habitat for Extreme Environment (SHEE). This is a European project. The goal is to develop a planetary habitat testbed for terrestrial analogue simulations [3].
- OEWF simulations: Analog field tests have been conducted by the Austrian Space Forum in different European locations. For instance, the Mars 2013 campaign was a 4-week Mars analog field test in Morocco held in February 2013 [5]. Experiments were carried out by international teams under simulated Martian surface exploration conditions. They were supervised by a Mission Support Center in Innsbruck, Austria.
- NSERC CREATE, Canadian Space Agency: Several experiments have been conducted in different Canadian locations to test robots and EVA.
- The Mars Yard. It is an Artificial Analogue of the Martian Surface located in Stevenage, UK.

Chapter 9

MISSION ARCHITECTURE OPTIONS AND ROADMAP

9.1. Main preparatory missions of the roadmap

The International Space Exploration Coordination Group (ISECG) has worked on a possible roadmap to the first human mission to Mars. A stepwise approach is proposed to develop key exploration technologies and capabilities. These steps include a mission to a Near-Earth object and a mission to the Moon.

Similar proposals can be found in the literature. However, there are still uncertainties on the configuration of the first human Mars mission, and that results in uncertainties in the roadmap.

For instance, if nuclear propulsion systems are used, additional milestones should be inserted in the roadmap. For architectures based on chemical propulsion, recent work suggests Heavy Mars Sample Return (HMSR) missions to qualify interplanetary vehicles, aerocapture, EDL systems, ISRU systems and Mars return. A synthesis of preparatory missions and their role is reported in Table 9.1.

All considered, for a given Mars mission architecture, the roadmap depends on complexity, costs, duration of the qualification phase, and the long-term sustainability of a Mars program. It is indeed difficult to justify a Mars flyby mission only from the technological and programmatic point of view. That said, according to some even a Mars flyby mission would help to convince public and politicians that a Mars mission is feasible. However, psychologically this may backfire or cause disaffection, and some astronauts have spoken against it.

Similar considerations apply for missions to the Moon. According to a recent study, for a class of simplified Mars architectures, a mission to the Moon before Mars may not be necessary or desirable. Nevertheless, for psychological reasons and in order to increase experience on human missions beyond LEO, it might be preferable to include a mission to the Moon in the roadmap even though it would have a significant impact on the cost and duration of the preparatory phase. Facing the follow-on costs associated to the ‘real’ Mars mission, politicians may decide to ‘postpone’ the Mars mission, with obvious consequences. A moon prepatory program could cause the Mars mission to come to a stand-still

Since costs and sustainability are key variables of the problem that are independent of scientific considerations, this Study Group was not able to find a consensus on the essential parameters of the mission and therefore on the roadmap. To clarify the reasons of this impasse, two mission approaches that have been discussed within the group are presented below.

The first approach tries to simplify the mission in order to minimize cost and duration of the preparatory phase; the second is based on a more ambitious mission requiring the development of new key technologies to facilitate interplanetary transportation.

	Earth	mission to ISS or LEO	Lunar vicinity, or HEO or NEO	Moon surface (human)	Mars vicinity (human)	Mars sample return (standard)	Heavy Mars sample return
Beyond low-Earth orbit crew transportation			++	++?	++		++
Heavy lift launch		++	++	++	++		++
Autonomous crew operations	+	++	++	++	++		
Deep space staging operations		+	++	++	++		++
Mars ascent	+			+		+	++
Space radiation protection /shielding	+	+	++	++	++		
Life support and habitation systems	+	++	++	++	++		
EDL systems	+			+		+	++
Surface power and energy management	++			++		+	+
Surface mobility	++			++		+	+
Human robotic integration	++	++	++	++	++		
Mars in-situ resource utilization	+					+	++
Long duration human health	+	++	++	++	++		
Deep space operations techniques	+	+	++	++	++	+	+
Mars Aerocapture						+	++
High velocity atmospheric Earth entry		+	++	++	++	+	++
Human factors	+	+	++	++	++		
Living in partial gravity	+	++	++	++	++		

Table 9.1. Possible preparatory missions and their benefits.
+: preliminary tests; ++: possible qualification

9.2. Simplified low cost missions

This class is defined by a small crew and several small interplanetary vehicles to be launched without LEO assembly. Main choices and features are in Table 9.2. One of the main constraints of this class of missions is Mars orbit aerocapture: it is assumed that all interplanetary vehicles are designed to enable safe aerocapture. One drawback of chemical propulsion is the difficulty in reducing Earth-Mars transit time, minimizing the impact of radiation dose. To compensate, this class of missions includes a fast free return trajectory as a mandatory backup option. This option needs a higher ΔV still considered manageable and the impact on IMLEO is not as high.

Remarkably, these simplifications would not be made at the expense of the risks. It is assumed in this class of missions that all systems are tested and qualified as usual.

Interestingly, with small Mars landing vehicles, the EDL procedures and systems would be simpler than those for heavier vehicles. Such simplifications would reduce the risks, facilitate the qualification phase (see Figure 9.1), and bring important costs reductions.

Mission parameter	**Choice**
Conjunction / opposition	Conjunction
Crew size	3 or 4
Interplanetary propulsion system	Chemical H_2/O_2
Mars orbit insertion	Aerocapture for all vehicles
Descent vehicles and EDL strategy	Small capsules <40 tons at Mars entry. Rigid heat shield or small HIAD.
ISRU options	O_2 or CH_4/O_2 produced on Mars
Launcher and LEO strategy	SLS class launcher, no need for LEO assembly
Vehicles configurations	Typically 3 or 4 rather small interplanetary vehicles to avoid LEO assembly (<130 tons in LEO and <40 tons at Mars entry); ERV in 2 modules that join in Mars orbit
Overall redundancy and multiple missions strategy	Total redundancy (e.g., 2 crews of 2) or redundancy of habitable module provided by cargo vehicle.
Main preparatory missions of the roadmap	2 main missions: Heavy Mars sample return mission High Earth orbit/ Moon orbit/ Mars flyby
Expected IMLEO	**≈ 500 tons (see [5])**

Table 9.2. Main choices for the class of simplified low cost missions.

An example of this simple mission class is in Figure 9.1 and a possible roadmap in Figure 9.2. The mission is novel and the roadmap is a direct illustration of an existing work. As the mission is simplified, an effort is also made for the roadmap, which is optimized for time and costs. The main idea is to qualify important systems including the interplanetary vehicles, aerocapture, and EDL systems thanks to HMSR. In order to achieve a high probability of success, at least two successful HMSR missions might be required. A complementary step could be achieved by means of several long-duration human missions to high Earth orbit, or eventually to the orbit of the Moon or the Lagrangian points. A crewed Mars flyby mission may also be appropriate, provided that the habitable module would be qualified first, which may imply preliminary long-duration human missions in Earth orbit.

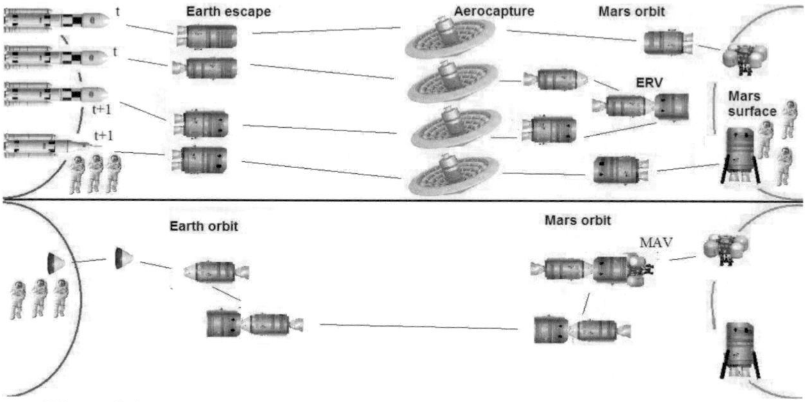

Figure 9.1: Possible scenario for a simplified low cost mission (original proposition).

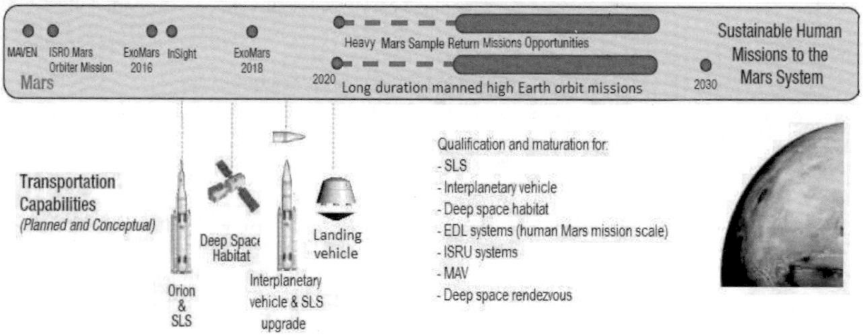

Figure 9.2: Simple roadmap for a low cost mission.

9.3. Missions with nuclear propulsion systems

This class of missions allow larger crews, heavier vehicles and the development of nuclear propulsion systems. There are two different technologies: nuclear thermal propulsion (NTP) and nuclear electric propulsion (NEP). NTP allows much higher thrust than NEP. For this reason, a NEP-powered mission would require power reactors in the 300 to 600 MW class. The Power/mass ratio of NEP is about 1/100 of that of NTP and NEP technologies may play an important role in the future (e.g., using a MagnetoPlasmaDynamic (MPD) thruster as in the VASIMR rocket), but their development and maturation might require many years. It could become feasible if a concurrent effort with mission planning were initiated. An option is to use NEP (or even solar electric propulsion) for cargoes not constrained by transit time but there would several different types of interplanetary propulsion systems in this case, which would increase the overall complexity of the mission. The tradeoff has been examined in the last NASA reference mission, where NTP is the preferred option for the three interplanetary vehicles.

Mission parameter	Choice
Conjunction / opposition	Conjunction
Crew size	5 to 6
Interplanetary propulsion system	NTP (preferred) or NEP
Mars orbit insertion	Aerocapture for cargos only (crew vehicle too big)
Descent vehicles and EDL strategy	Heavy landers. Large HIAD.
ISRU options	O_2 or CH_4/O_2 produced on Mars
Launcher and LEO strategy	SLS class launcher, long LEO or MEO assembly
Vehicles configurations	Typically 3 interplanetary vehicles: 1 cargo with MAV; 1 cargo with Mars surface habitat; 1 crewed interplanetary vehicle
Overall redundancy and multiple missions strategy	Redundancy with assets of next mission
Main preparatory missions of the roadmap	Human mission to Moon orbit Human mission to the surface of the Moon Human mission to the asteroids Mars sample return mission EDL tests in Mars atmosphere
Expected IMLEO	**≈ 900 tons (see [21])**

Table 9.3. Main choices for the class of missions with nuclear propulsion systems (like NASA reference mission).

The advantage of this NASA-like mission is a reduced IMLEO in comparison with the chemical propulsion option (assuming the same number of astronauts, no aerocapture for the crewed vehicle, same number and configuration of vehicles, and same scenario).

It is also an important step in the development of nuclear technologies in space, as they will be developed anyway. Another long-term advantage of NTP or NEP is the possibility of re-using the same vehicle for several missions, provided maintenance is feasible and affordable.

The main drawbacks of this class of missions are the duration of the preparatory phase for the qualification of all systems, the cost and the complexity of the roadmap.

As suggested by ISECG, a mission to the surface of the Moon would help maturing the technologies and procedures regarding human space missions. In addition, since long periods of time are required for the developments and tests of new systems (nuclear technologies, EDL systems for heavy landers, etc.), such missions also are important to carry on experiencing on human space missions and to avoid losing the basic knowledge concerning the manufacturing and exploitation of these systems.

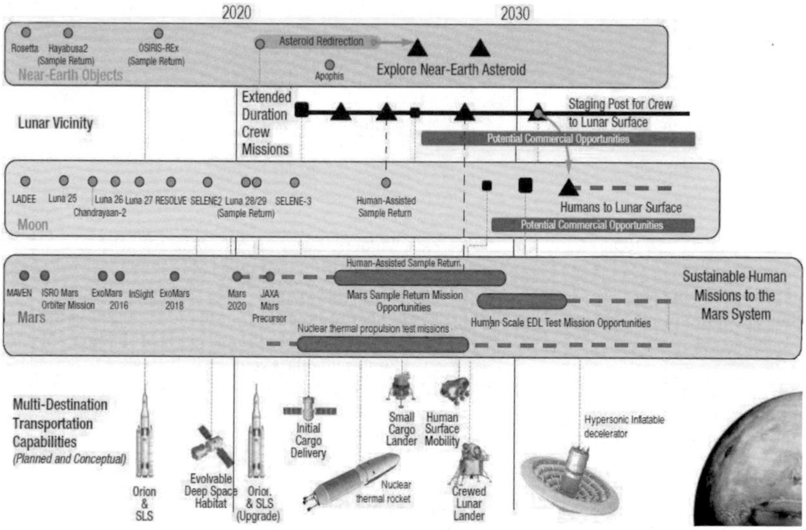

Figure 9.3: Roadmap with NTP (based on ISECG's work and complementary missions for NTP qualification)

Chapter 10
CONCLUSIONS

The human exploration of Mars is considered – in the short term – the final goal of human space exploration. In a longer perspective it is a vital step to create a spacefaring civilization.

In recent years, the robotic exploration of Mars has had outstanding successes. We now know much more about the planet than we did just a few years ago. This knowledge helps along the way to exploration. Many other steps must follow to proceed along this path, as shown by the ISEGC exploration roadmap.

Sending humans to Mars, giving them all that is needed to perform their tasks on the planet, and bringing them safely home is a formidable task which requires a long-term engagement of a number of Countries working together in what will be the most complex and daring collective effort ever performed by an international group in times of peace.

Like in all other space missions, the first critical step is reaching Earth orbit. Any decrease of cost in this phase may make the very large IMLEO involved in Mars missions more acceptable. Availability of heavy launchers will reduce the number of launches and the need for assembling interplanetary vehicles from predeployed payloads. From this viewpoint, the production of propellant on the Moon or in space would be of great help, but the relevant technologies still lie in an uncertain future.

Human factors are critical to mission success, most especially those linked to radiation exposure and long stays in microgravity. Further studies should find solutions for both. Until passive or active radiation shielding technologies are developed, fast transit to Mars and back seems to be the only solution to reducing the radiation dose. The second, slightly less critical, concern is the effect of exposure to microgravity has on the crew. Here there are solutions other than a fast journey, such as exercising and, above all, creating an artificial gravity by rotating the either the whole spacecraft or a part of it.

Psychological factors are critical and should be given due concideration. They are crucial in deciding the crew size, a factor impacting the whole mission architecture.

Among the most critical technical factors are those relating to deep space propulsion. The choice is between conventional approaches, with high TRL and lower cost, but characterized by lower performance, and more advanced solutions, requiring more R&D investments but allowing faster transit and opening the way to future developments.

It is important that those who propose a human mission to Mars realize that humankind cannot afford a false start in its way to Mars, and affordability and safety must be taken into consideration since the beginning. A single accident, especially if it leads to the loss of the mission or, even worse, the loss of the crew. Even minor setbacks like unexpected cost growth, large delays, or withdrawal of important partners may cause the program to be discontinued, a setback which could last decades. The choice of the relevant technologies, accurately balancing performance with technology readiness, and safety must be done conscious that the first mission to Mars must not be just a flag-and-footprint mission or the final point of an ancient dream, but a mission which will open a new era of exploration, aimed at transforming humankind into a spacefaring species.

Chapter 11
RECOMMENDATIONS

As a conclusion of the study, the study group formulates the following recommendations:

1) As no consensus was found on some important technological issues, carry on working on the details of each scenario, try to determine the costs and the most efficient roadmap for each option and provide all relevant information for decision makers for the final choice.
2) Advise in defining an International Mars reference mission scenario, with the involvement of Space Agencies and Industries, to agree on a preliminary technical baseline and the required technological decision milestones, for instance:
   a. Nuclear thermal and nuclear electric propulsion.
   b. Zero-boil off technology for cryogenic propellant storage.
   c. Light structures and heavy launchers.
   d. Nuclear power generator systems for both space and on-planet usage.
   e. Passive or active radiation shielding technology.
   f. Artificial gravity in space.
   g. Effects of Mars gravity using large centrifuges or tethered spacecraft in orbit.
   h. Aerocapture technologies for large payloads.
   i. Life support systems, particularly regenerative ones.
   j. ISRU systems.
   k. Exploration technologies, e.g. astronaut robotic assistants, rovers, drillers, etc.
3) Define and implement the Human Mars Mission Feasibility Index (HMMFI) (see Appendix D).
4) Set up a joint working group, IAA (SG3.16)/ISECG, to define in which of the Human Missions beyond Earth, defined in the ISECG roadmap, the technologies defined in the Mars reference mission (see recommendations 1) should be demonstrated, in order to reduce the risk and cost of the Global Mission to Mars. Demonstration projects, to be carried out by a variable group of countries will be defined to this end.
5) Foster the global involvement of countries, particularly the emerging and developing countries, through existing bodies like ISECG, UNOOSA-HSTI, IAA, etc.
6) Improve the common knowledge of the human factors, as a critical issue for human Mars missions.
7) Make use of the Human Spaceflight Virtual Institute, by IAA and, with the participation of various space agencies and industries, foster the exploitation of existing technologies, facilities and knowledge available world-wide. This Institute will also facilitate the engagement of new and developing countries by identifying technological niches existing in these countries, as well as facilitating the exchange of information in many critical areas, such as human factors.

The institution of a follow-up Study group dealing with specific issues mentioned above is encouraged.

REFERENCES

Chapter 1.

References on Mars mission architectures:

[1]. J.E. Brandenburg, J. Kline, R. Cohen, and K. Diamante, MarsX : Mars in 10 years, Proc. of the 5th Mars Society Convention, Boulder, Colorado, 2002.

[2]. John Brandenburg, Mars-Phobos: A Mars Mission Architecture with Mars-Moon Synergy, proc. of the Space Technology & Applications International Forum (STAIF II), 12-13 March 2012.

[3]. E. Cliquet, R. Heidmann, A. Souchier, Manned Mars Missions Scenarii and Highlights on Architectures, 42nd AIAA/ASME/SAE/ASEE Joint Propulsion Conference & Exhibit, AIAA 2006-5126, Sacramento, California (USA), 9-12 July 2006.

[4]. K. Sankaran, L. Cassady, A.D. Kodys and E.Y. Choueiri, A Survey of Propulsion Options for Cargo and Piloted Missions to Mars, International Conference on New Trends in Astrodynamics, January 20-22, 2003.

[5]. T.D. Schmidt, W. Seboldt and M. Auweter-Kurtz, Flexible piloted Mars mission using continuous electric propulsion, Journal of Spacecraft and Rockets, vol. 43 (6), 2006.

[6]. R. Myers, C. Carpenter, High Power Solar Electric Propulsion for Human Space Exploration Architectures, proc. of the 32nd International Electric Propulsion Conference, Wiesbaden, Germany, sept. 11-15, 2011.

[7]. M. Raftery, D. Cooke, J. Hopkins, B. Hufenbach, An affordable mission to Mars, proceedings of the 64th Int. Astronautical Conference, IAC-13, A5, 4-D2.8.4, Beijing (China), October 2013.

[8]. A.V. Ilin, L.D. Cassady, T.W. Glover, M.D. Carter and F.R. Chang Diaz, A Survey of Missions using VASIMR for Flexible Space Exploration, JSC report 65825, 2010.

[9]. L. Bessone and D. Vennemann, CDF Study Report, Human Missions to Mars, Overall Architecture Assessment, CDF-20(A), February 2004.

[10].J. Christian, C. Tanner and J. Theisinger, A Revolutionary Outpost Architecture for Human Mars Exploration, proc. of the IAC, IAC-06-D1.1.7, 2006.

[11].International Academy of Astronautics, The International Exploration of Mars, 4th Cosmic Study of the IAA, April 1993.

[12].International Academy of Astronautics, The Next Steps in Exploring Deep Space, W. Huntress ed., July 2004.

[13].W. W. Mendell and V. Legostaev, International Mars Mission, Final Report of the 1991 ISU Design Project, International Space University, Cambridge, MA, USA, 1991.

[14].G. Bonin, Reaching Mars for Less: The Reference Mission Design of the MarsDrive Consortium, 25th International Space Development Conference (ISDC), Los Angeles, California, 4-7 May 2006.

[15].D.Willson et J.D.A Clarke, A Practical Architecture for Exploration-Focused Manned Mars Missions Using Chemical Propulsion, Solar Power Generation and In-

Situ Resource Utilisation, Proceedings of the 6th Australian Space Science Conference, p. 186-211, 2006.
[16].C. Hirata, N. Brown, D. Shannon, J.D. Burke, B. Murray, M. Adler (The Mars Society at California Institute of Technology), Mars SCHEME, 3rd Annual HEDS-UP Forum, May 4–5, 2000.
[17].Report of the 90-Day Study on Human Exploration of the Moon and Mars, NASA, November 1989.
[18].S. J. Hoffman and D. I. Kaplan, eds., Human Exploration of Mars: The Reference Mission of the NASA Mars Exploration Study Team - NASA SP-6107 - NASA Johnson Space Center, July 1997.
[19].B. G. Drake, ed., Reference Mission Version 3.0 Addendum to the Human Exploration of Mars: The Reference Mission of the NASA Mars Exploration Study Team - EX13-98-036 - Exploration Office, NASA Johnson Space Center, June, 1998.
[20].S. K. Borowski and L. A. Dudzinski, Vehicle and Mission Design Options for the Human Exploration of Mars/Phobos using "Bimodal" NTR and LANTR propulsion, 34th Joint Propulsion Conference of the American Institute of Aeronautics and Astronautics, Cleveland, Ohio, 13-15 juillet 1998.
[21].B.G. Drake ed., Mars Architecture Steering Group, Human Exploration of Mars, Design Reference Architecture 5.0 (and addendums), NASA Johnson Space Center, 2009.
[22].S. K. Borowski, D. R. McCurdy and T. W. Packard, "7-Launch" NTR Space Transportation System for NASA's Mars Design Reference Architecture (DRA) 5.0, 45th Joint Propulsion Conference of the American Institute of Aeronautics and Astronautics, August 2-5, 2009.
[23].Hawkins and T. Radcliffe, Austere Human Missions to Mars, AIAA SPACE 2009 Conference, Pasadena, 14-17 septembre 2009.
[24].Mars Architecture Steering Group, Addendum to the Human Exploration of Mars, Design Reference Architecture 5.0, NASA Johnson Space Center, 2014.
[25].D. Portree, Humans to Mars: Fifty Years of Mission Planning, 1950 - 2000, NASA Monographs in Aerospace History Series, no 21, February 2001.
[26].J.M. Salotti, Simplified scenario for manned Mars missions, Acta Astronautica, vol. 69, 266–279, 2011.
[27].J.M. Salotti, Revised Scenario for Human Missions to Mars, Acta Astronautica, vol. 81, p. 273-287, 2012.
[28].B. St Germain, J.R. Olds, J. Bradford, A.C. Charania, D. DePasquale, M. Schaffer, J. Wallace, Uilizing Lunar Architecture Transportation Elements for Mars Exploration, proc. of the AIAA Space 2007 Conference, Long Beach, California, 18-20 septembre 2007.
[29].Tavana, M., Zandi, F., Applying fuzzy bi-dimensional scenario-based model to the assessment of Mars mission architecture scenarios, Advances in Space Research, volume 49, issue 4, pp. 629 – 647, 2012.
[30].W. von Braun, Manned Mars Landing, Presentation to the Space Task Group, NASA document, presentation materials, 4th August 1969.
[31].D.B. Weaver and M.B. Duke, Mars Exploration Strategies: A Reference Program and Comparison of Alternative Archi-tectures, AIAA 93-4212, AIAA Space Programs and Technology Conference, September 21-23, 1993.
[32].D.B. Weaver and M.B. Duke, Mars Exploration Strategies: A Reference Design Mission, proc. of the IAF 93-Q.1.383, 1993.

[33].R. Zubrin and D. A. Baker, Mars Direct: Humans to the Red Planet by 1999, proceedings of the 41st Congress of the International Astronautical Federation, 1990.
[34].R. Zubrin and D. Weaver, Practical Methods for Near-Term Piloted Mars Missions, AIAA 93-2089 AIAA/SAE 29th Joint Propulsion Conference, Monterey CA, 1993.

Other references:

[35].N. Augustine, Review of Human Spaceflights Plan Committee, report of the Advisory Committee on the Future of the U.S. Space Program, October 2009.
[36].R.D. Braun and R.M. Manning, Mars Entry, Descent and Landing Challenges, Journal of Spacecraft and Rockets, Vol. 44 (2), 310–323, Mar-Apr, 2007.
[37].G. Cates, C. Stromgren, D. Arney, K. Goodliff, International Human Mission to Mars: Analyzing a Conceptual Launch and Assembly Campaign, proceedings of the IEEE Aerospace Conference, ref. 8.0106, Big Sky, Montana (USA), March 2014.
[38].J.A. Christian, K. Manyapu, G.W. Wells, J.M. Lafleur, A.M. Verges and R.D Braun, Sizing of an Entry, Descent, and Landing System for Human Mars Exploration, Journal of Spacecraft and Rockets, Vol. 45, No. 1, pp. 130–141, Jan-Feb, 2008.
[39].Y. Clearwater, Human Factors and Crew Size Studies, Presentation at Mars Exploration Study Workshop II, NASA Conference Publication 3243, NASA Ames Research Center, May 24-25, 1993.
[40].C. Cooper, W. Hofstetter, J. A. Hoffman, E.F. Crawley., Assessment of architectural options for surface power generation and energy storage on human Mars missions, Acta Astronautica, vol. 7-8, 1106-112, 2010.
[41].D. Crispa, A. Pathareb, R.C. Ewella, The performance of gallium arsenide/germanium solar cells at the Martian surface, Acta Astronautica vol. 54, 83-101, 2003.
[42].ESAS report, Chapter 8: Risk and Reliability, NASA's Exploration Systems Architecture Study, technical report NASA-TM-2005-214062, November 2005.
[43].Hufenbach, K. C. Laurini, J.-C. Piedboeuf, B. Schade, K. Matsumoto, F. Spiero, A. Lorenzoni, The Global Exploration Roadmap, proceedings of the 62nd Int. Astr. Congress, IAC-11.B3.1.8, Cape Town, October 2011.
[44].International Space Exploration Coordinating Group, The Global Exploration Roadmap, ISECG Technical Report, August 2013.
[45].A.M. Kurzun, G. F. Dubos, C.K. Iwata, B.A. Stahl, A concept for the entry, descent, and landing of high-mass payloads at Mars, Acta Astronautica, vol. 66, p. 1146-1159, 2010.
[46].A.T. May, S.D. Creech, NASA's Space Launch System (SLS) Program: Mars Program Utilization, NASA MSC report no 20120015287, 2012.
[47].J.M. Salotti and B. Claverie, Human Mission to Mars: All-up vs Pre-deploy, proceedings of the Global Space Exploration Conference, GLEX-2012-08.2.7, Washington D.C., 22-24 May 2012.
[48].J.M. Salotti and E. Suhir, Manned Missions to Mars: Minimizing Risks of Failure, Acta Astronautica, vol. 93, p. 148–161, 2014.
[49].J.M. Salotti and E. Suhir, Some Major Guiding Principles to Make Future Manned Missions to Mars Safe and Reliable, Proceedings of the IEEE Aerospace Conference, Big Sky, Montana (USA), pp. 1-6, March 2014.
[50].J.M. Salotti, R. Heidmann and E. Suhir, Crew Size Impact on the Design, Risks and Cost of a Human Mission to Mars, Proceedings of the IEEE Aerospace Conference, Big Sky, Montana (USA), pp. 1-9, March 2014.

[51].J.M. Salotti and R. Heidmann, Roadmap to Human Mars Mission, Acta Astronautica, accepted (accepted 3rd of June 2014).
[52].D. Tito, G. Anderson, J.P. Carrico Jr, J. Clark, B. Finger, G.A. Lantz, M.E. Loucks, T. MacCallum, J. Poynter, T.H. Squire and S.P. Worden, Feasibility Analysis for a Manned Mars Free-Return Mission in 2018, Proceedings of the IEEE Aerospace Conference, Big Sky, Montana (USA), pp. 1-18, March 2013.
[53].M. A. Viscio, E. Gargioli, J.A. Hoffman, P. Maggiore, A. Messidoro and N. Viola, A methodology to support strategic decisions in future human space exploration: From scenario definition to building blocks assessment, Acta Astronautica, vol. 91, 198-217, October–November 2013.

Chapter 2.

[1] Von Braun, *Das Mars Projekt*, 1949

Chapter 3.

[1] National Research Council, Approaches to Future Space Cooperation and Competition in a Globalizing World: Summary of a Workshop, Washington, DC: The National Academies Press, 2009
[2] The International Exploration of Mars, IAA Mars Cosmic Study Report, 4th Cosmic Study of the IAA, International Academy of Astronautics, 1996
[3] The Global Exploration Strategy: the Framework for Coordination, April, 2007
[4] Review of US Human Spaceflight Plans Committee, Seeking a Human Spaceflight Program Worthy of a Great Nation, Report to White House Office of Science and Technology Policy, October 2009
[5] Human Space Exploration Framework Summary, NASA, January 2010, http://www.nasa.gov/pdf/509813main_Human_Space_Exploration_Framework_Summary-2010-01-11.pdf
[6] D. A. Broniatowski et al., " The Case for Managed International Cooperation in Space Exploration" International Space Exploration Update, Center for Strategic and International Studies
[7] Space Studies Board, Assessment of Impediments to Interagency Collaboration on Space and Earth Science Missions. National Academies Press, 2011.
[8] J.D. Rendleman, Perspectives on Improving United States International Space Cooperation
[9] Statement of Steven W. Squyres, Before the Subcommittee on Space, United States House of Representatives, June 19, 2013
[10] European Space Policy Progress Report, Commission of the European Communities, COM(2008)561 final, Brussels, November 9, 2008
[11] European Space Agency, Status Future Plans and Rationale for Cooperation, 2nd ASI/ESA Workshop on International Cooperation for Sustainable Space Exploration, Sarteano, Italy, May 2006
[12] D. Sacotte, Rationale, Philosophy and Objectives of the Workshop - The Workshop in the Context of the European Aurora Programme, 2nd ASI/ESA Workshop on

International Cooperation for Sustainable Space Exploration, Sarteano, Italy, May 2006
[13] Many Benefits from International Cooperation, But Not Cost Savings, Says Panel, Laura Delgado, Sace News, November 20, 2013
[14] J.V. Zimmerman, The IAF and Space Exploration, 2nd ASI/ESA Workshop on International Cooperation for Sustainable Space Exploration, Sarteano, Italy, May 2006
[15] *The Global Exploration Roadmap*, International Space Exploration Coordination Group, August 2013
[16] International Space Exploration Coordination Group, Terms of Reference
[17] National Space Policy of the United States of America, June 28, 2010
[18] J.-J. Dordain, "Above and Beyond", remarks at NASA's Project Management Challenge, February 9. 2010, Galveston, Texas.
[19] *International Space Station Lessons Learned As Applied to Exploration*, Kennedy Space Center, July 22, 2009
[20] K. Laurini et al., International Space Station (ISS) Lessons Learned and Their Influence on Preparations for Human Exploration Beyond Low Earth Orbit, 62nd International Astronautical Congress, Cape Town, SA, 2011
[21] Report of Session 5: International Cooperation of 1st Workshop on International Cooperation for Sustainable Space Exploration, Sarteano, Italy, 3 – 6 May 2005
[22] D. Southwood, Research collaboration: When international partnerships go wrong' *Nature,* 488, 451–453, August 23, 2012
[23] C. Bank and M. Caporicci, EU-BRIC Cooperation in Space Example: "Crew Space Transportation System", Symposium on Legal and Policy Aspects of Space Cooperation between Europe and BRICS Countries: Inventory, Challenges and Opportunities, Leiden University, May 6, 2013
[24] J. Hopkins, An International Industry Perspective on Extended Duration Missions Near the Moon, NASA Global Exploration Roadmap Community Workshop, Laurel, Maryland, April 10, 2014
[25] G. Genta, *Private Space Exploration: a New Way for Starting a Spacefaring Society?* Acta Astronautica, Vol. 104, Apr. 2014, pp. 480-486.
[26] S. Stich, Asteroid Redirect Mission and Human Space Flight Briefing to National Research Council Committee for Study on Human Space Flight Technical Panel, June 19, 2013

Chapter 4.

[1]. Genta G. *Introduction to the Mechanics of Space Robots*, Springer, New York, 2011
[2]. U.S. Extension to the ICAO Standard Atmosphere, U.S. Government Printing Office, Washington, D.C., 1958.
[3]. U.S. Standard Atmosphere, 1976, U.S. Government Printing Office, Washington, D.C., 1976
[4]. Kenneth R. Lang, *The Sun from Space,* Springer, New York, 2000.
[5]. Bruce Jakosky, *The Search for Life on Other Planets*, Cambridge University Press, Cambridge, 1998.

[6]. Laurence Bergreen, *The Quest for Mars*, Harper Collins, London, 2000.
[7]. http://www.nasa.gov/pdf/373665main_NASA-SP-2009-566.pdf.
[8]. http://www.nasa.gov/pdf/373667main_NASA-SP-2009-566-ADD.pdf
[9]. Christopher P. McKay Planetary Ecosynthesis on Mars: Restoration Ecology and Environmental Ethics NASA Ames Research Center.
[10].R. Zubrin, R. Wagner, *The Case for Mars,* Touchstone, New York, 1997, and R. Zubrin, *Entering Space*, Tarcher/Putnam, New York, 2000.
[11].Beaty, D.W.; Buxbaum, K.L.; Meyer, M.A.; Barlow, N.G.; Boynton, W.V.; Clark, B.C.; Deming, J.W.; Doran, P.T.;Edgett, K.S.; Hancock, S.L.; Head, J.W.; Hecht, M.H.; Hipkin, V.; Kieft, T.L.; Mancinelli, R.L.;McDonald, E.V.; McKay, C.P.; Mellon, M.T.; Newsom, H.; Ori, G.G.; Paige, D.A.; Schuerger, A.C.;Sogin, M.L.; Spry, J.A.; Steele, A.; Tanaka, K.L.; Voytek, M.A.; (2006).Findings of the Mars Special Regions Science Analysis Group, Unpublished white paper, 76 p, posted June 2006 by the Mars Exploration Program Analysis Group (MEPAG) at http://mepag. jpl.nasa.gov/reports/index.html.
[12].Kminek, G., J. D. Rummel, C. S. Cockell, R. Atlas, N. Barlow, D. Beaty, W. Boynton, M. Carr, S. Clifford, C. A. Conley, A. F. Davila, A. Debus, P. Doran, M. Hecht, J. Heldmann, J. Helbert, V. Hipkin, G. Horneck, T. L. Kieft, G. Klingelhoefer, M. Meyer, H. Newsom, G. G. Ori, J. Parnell, D. Prieur, F. Raulin, D. Schulze-Makuch, J. A. Spry, P. E. Stabekis, E. Stackebrandt, J. Vago, M. Viso, M. Voytek, L. Wells, F. Westall, and C. Cockell, Report of the COSPAR Mars Special Regions Colloquium, *Adv. Space Res*.**, 46,** (No. 6). 811-829, 2010.
[13].Rummel, J. D., D. W. Beaty, M. A. Jones, C. Bakermans, N. G. Barlow, P. J. Boston, V. F. Chevrier, B. C. Clark, J.-P. P de Vera, R. V. Gough, J. E. Hallsworth, J. W. Head, V. J. Hipkin T. L. Kieft, A. S. McEwen, M. T. Mellon, J. A. Mikucki , W. L. Nicholson, C. R. Omelon, R. R. Peterson, E. Eric, L. B. Sherwood , K. L. Tanaka., D. Viola and J. J. Wray, A new analysis of Mars Special Regions :Findings of the second MEPAG Special Regions Science Analysis Group *SR-SAG2 Astrobiology*. **14**(11): 887-968. doi:10.1089/ast.2014.1227, 2014.
[14].Summons, R.E., A. L. Sessions, A. C. Allwood, H. A. Barton, D. W. Beaty, B. Blakkolb, J. Canham, B. C. Clark, J. P. Dworkin, Y, Lin, R. Mathies, A. M. Milkovich and A. Steele, Planning considerations related to the organic contamination of martian samples and implications for the Mars 2020 rover. *Astrobiology* 14 (11), doi:10.1089/ast.2014.1405. 2014.

Chapter 5.

[1]. Charles, J.B. "Space station, lunar base, on to Mars. Human health and performance aspects of early interplanetary expeditions." Presentation material, April 2008.
[2]. Jones, J.A., Barratt, M., Effenhauser, R., Cockell, C.S., Lee, P. (2004). Medical issues for a human mission to Mars and martian surface expeditions." J. Brit.Interplanetary Soc., 57, 144-160.
[3]. Adams, J. H. Jnr., Silberberg R. and Tsao, C. H., (198P1). Cosmic Ray Effects on Microelectronics Part I. The Near Earth Particle Environment, NRL Memo Rep., 4506 Pt. I. Publ. U.S. Navy, August 1981.

[4]. McKenna-Lawlor, S. (2016). An overview of the energetic particle radiation hazard en route to and at Mars in 'The Particle Radiation Hazard en route to and at Mars' (Publ. Internat. Acad. Astronautics), in press.
[5]. McKenna-Lawlor, S., Gonçalves, P., Reitz, G., and Matthia, D. (2011). Overview of energetic particle hazards during prospective manned missions to Mars. Planet. Space Sci., doi:10.1016/j.pss.2011. 06.017.
[6]. McKenna-Lawlor, S., Gonçalves, P.; Keating, A., Morgado, B., Heynderickx, D., Nieminen, P., Santin, G., Truscott, P., Lei, F., Foing, B. and Balaz, J. (2012). Characterization of the particle radiation environment at three potential landing sites on Mars using ESA's MEREM models, Icarus 218, 723-734.
[7]. Townsend, L. W., Pour Arsalan, M., Hall, M.I., Anderson, J.A., Bhatt, S., DeLauder N., Adamczyk, A.M. (2011). Estimates of Carrington-class solar particle event radiation exposures on Mars, Acta Astron., 69, 397-405.
[8]. McKenna-Lawlor, S. and the SG.3.19/1.10 team (2014), Feasibility study of astronaut standardized career dose limits in LEO and the outlook for BLEO, Acta Astron. 104, 565-573.
[9]. Zeitlin, C. J., D. M. Hassler, and R. F. Wimmer-Schweingruber (2012). MSL RAD dosimetric measurements in Cruise and on Mars. Calibration and first results, AGU Fall Meeting Paper SH21A-2196.
[10].Hassler, D. M., C. Zeitlin, R. F. Wimmer-Schweingruber, B. Ehresmann, S. Rafkin, J. L. Eigenbrode, D. E. Brinza, G. Weigle, S. Böttcher, E. Böhm, S. Burmeister, J. Guo, J. Köhler, C. Martin, G. Reitz, F. A. Cucinotta, M.-H. Kim, D. Grinspoon, M. A. Bullock, A. Posner, J. Gómez-Elvira, A. Vasavada and J. P. Grotzinger,(2013). Mars' Surface Radiation Environment Measured with the Mars Science Laboratory's Curiosity Rover. *Science*, doi: 10.1126/science.1244797.
[11].Cucinotta, F. A., Schimmerling, W., Wilson, J.W., Peterson, L.E., Badhwar, G.D.,Saganti, P.B., Dicullo, J.F. (2001a). Space radiation cancer risk projections for exploration missions: uncertainty reduction and mitigation. NASA/TP, JSC-29295 (Publ. Johnson Space Centre, USA).
[12].Cucinotta, F. A., Schimmerling, W., Wilson, J.W., Peterson, L.E., Badhwar, G.D., Saganti, P.B., Dicullo, J.F. (2001b).Space radiation cancer risks and uncertainties for Mars missions. Radiat. Meas., 156, pp. 682–685.
[13].Reitz, G., (2016). Radiation Health impacts on the way to Mars, in 'The Particle Radiation Hazard en route to and at Mars' (Publ. Internat. Acad. Astronautics), in press.
[14]. Clancy, P., A. Brack and G. Hornick ((2006) in Looking for life. Search the Solar System (Publ. Cambridge Univ. Press).
[15].McKenna-Lawlor, S. and the SG.3.19/1.10 team (2015), Recommendations to mitigate against human health risks incurred due to energetic particle irradiation Beyond Low Earth Orbit (BLEO), Acta Astron. 109, 182-193.
[16].Newberg A.B. (1994). Changes in the central nervous system and their clinical correlates during long-term spaceflight. Aviat., Space, Environ. Med., 66, 562-572.
[17].Kozlovskaya I.B., Kreidich Y.V., Oganov V.S., Koserenko O.P. (1981). Pathophysiology of motor functions in prolonged manned space flights. Acta Astronaut., 8, 1059-1072.

[18].Grigoriev A.J., Polyakov V.V., Bogomolov V.V., Egorov A.D., Pestov I.D., Kozlovskaya I.B. (1990). Medical Results of the Fourth Prime Expedition on the Orbital Station MIR. Proc. of the Fourth European Symposium on Life Sciences Research in Space, Trieste, ESA SP-307: 19-22.

[19].Berger M., Mescheriakov S., Molokanova E., Lechner-Steinleitner S., Seguer N. & Kozlovskaya I. (1997). Pointing ar.m movements in short- and long-term spaceflights. Aviat., Space, Environ. Med., 68, 781-787

[20].Bock O., Fowler B. & Comfort D. (2001). Human sensorimotor coordination during spaceflight: An analysis of pointing and tracking responses during the "Neurolab" Space Shuttle mission. Aviat., Space, Environ. Med., 72, 877-883.

[21].Heuer H., Manzey D., Lorenz B., Sangals J. (2003). Impairments of manual tracking performance during spaceflight are associated with specific effects of microgravity on visuomotor transformations. Ergonomics, 46, 920-934.

[22].Watt D.G.D. (1997). Pointing at memorized targets during prolonged microgravity. Aviat., Space, Environ. Med., 68, 99-103.

[23].Semjen A., Leone G., Lipshits M. (1998). Motor timing under microgravity. Acta Astronaut., 42, 303-321.

[24].Pattyn N., Migeotte P., Demaesselaer W., Kolinsky R., Morais J., Zizi M. (2005). Investigating human cognitive performance during spaceflight. J. Gravit. Physiol., 12, P39 – P40.

[25].Lackner J.R. (1985). Human sensory-motor adaptation to the terrestrial force environment. In: D. Ingle (ed.)., Brain Mechanisms and Spatial Vision, vol. 21. Dordrecht, The Netherlands: Springer.

[26].Roll J.P., Popov K., Gurfinkel V., Lipshits M., Andre-Deshays C., Gilhodes J.C., Quoniam C. (1993). Sensorimotor and perceptual function of muscle proprioception in microgravity. J. Vestibular Res., 3, 259-273.

[27].Manzey D., Lorenz B. (1998) Mental performance during short-term and long-term spaceflight. Brain Res. Rev., 28, 215-221.

[28].Dinges, D. F., Powell, J.W. (1985). Microcomputer analyses of performance on a portable, simple visual RT task during sustained operations. Behavior Research Methods, Instruments, & Computers, 17, 652-655.

[29].Kanas, N., Manzey, D. (2008). Space Psychology and Psychiatry, 2nd Edition. El Segundo, California: Microcosm Press; and Dordrecht, The Netherlands: Springer.

[30].Kanas, N., Salnitskiy, V., Grund, E.M., Gushin, V., Weiss, D.S., Kozerenko, O., Sled, A., Marmar, C.R. (2000). Interpersonal and cultural issues involving crews and ground personnel during Shuttle/Mir space missions. Aviat., Space, Environ. Med., 71 (9, Suppl.), A11-16.

[31].Kanas, N., Salnitskiy, V., Weiss, D.S., Grund, E.M., Gushin, V., Kozerenko, O., Sled, A., Bostrom, A., Marmar, C.R. (2001). Crewmember and ground personnel interactions over time during Shuttle/Mir space missions. Aviat., Space, Environ. Med., 72, 453-461.

[32].Kanas, N.A., Salnitskiy, V.P., Boyd, J.E., Gushin, V.I., Weiss, D.S., Saylor, S.A., Kozerenko, O.P., & Marmar, C.R. (2007). Crewmember and mission control personnel interactions during International Station Missions. Aviat., Space, Environ. Med., 78, 601-607.

[33].Boyd, J.E., Kanas, N.A., Salnitskiy, V.P., Gushin, V.I., Saylor, S.A., Weiss, D.S., Marmar, C.R. (2009). Cultural differences in crewmembers and mission control personnel during two space station programs. Aviat., Space, Environ. Med., 80, 1-9.

[34].Bechtel, R.B., Berning, A. (1991). The third-quarter phenomenon: Do people experience discomfort after stress has passed? In: A.A. Harrison, Y.A. Clearwater, C.P., McKay, eds., From Antarctica to Outer Space. New York: Springer-Verlag.

[35].Gushin, V.I., Zaprisa, N.S., Kolinitchenko, T.B., Efimov, V.A., Smirnova, T.M., Vinokhodova, A.G., Kanas, N. (1997). Content analysis of the crew communication with external communicants under prolonged isolation. Aviat., Space, Environ. Med., 68, 1093-1098.

[36].Gushin, V.I. (2003). Problems of distant communication of isolated small groups. Human Physiol., 29, 548-555.

[37].Gushin. V.I., Pustynnikova, J.M., Smirnova, T.M. (2001). Interrelations between the small isolated groups with homogeneous and heterogeneous composition. Human Perf. in Extreme Environ., 6, 26-33.

[38].Gushin, V.I., Efimov, V.A., Smirnova, T.M., Vinokhodova, A.G, Kanas, N. (1998). Subject's perception of the crew interaction dynamics under prolonged isolation. Aviat., Space, Environ. Med., 69, 556-561.

[39].Vinokhodova, A.G., Gushin, V.I. (2012). Study of values and interpersonal perception in cosmonauts on board of International Space Station. Paper # IAC-12-A1.1.8. International Astronautical Federation. Proceedings, 63th International Astronautical Congress, Naples, Italy, October 1-5, 2012.

[40].Tomi, L., Kealey, D., Lange, M., Stefanowska, P., Doyle, V. (2007). Cross-cultural training requirements for long-duration space missions: Results of a survey of International Space Station astronauts and ground support personnel. Paper delivered at the Human Interactions in Space Symposium, May 21, 2007, Beijing, China.

[41].Nechaev, A.P., Polyakov, V.V., Morukov, B.V. (2007). Martian manned mission: What cosmonauts think about this. Acta Astronaut., 60, 351-353.

[42].Sandal, G.M., Manzey, D. (2009). Cross-cultural issues in space operations: A survey study among ground personnel of the European Space Agency. Acta Astronaut., 65, 1520-1529.

[43].Stuster, J. (2010). Behavioral Issues Associated with Long-Duration Space Expeditions: Review and Analysis of Astronaut Journals. Experiment 01-E104 (Journals): Final Report. NASA/TM-2010-216130. Houston, Texas: NASA/Johnson Space Center.

[44].Urbina, D., Charles, R. (2012). Enduring the isolation of interplanetary travel: A personal account of the Mars 500 mission. Paper # IAC-12-A1.1.1. International Astronautical Federation. Proceedings, 63th International Astronautical Congress, Naples, Italy, October 1-5, 2012.

[45].Kanas, N., Saylor, S., Harris, M., Neylan, T., Boyd, J., Weiss, D.S., Baskin, P., Cook, C., Marmar, C. (2010). High vs. low crewmember autonomy in space simulation environments. Acta Astronaut., 67, 731-738.

[46].Kanas, N., Harris, M., Neylan, T., Boyd, J., Weiss, D.S., Cook, C., Saylor, S. (2011). High vs. low crewmember autonomy during a 105-day Mars simulation mission. Acta Astronaut., 69, 240-244.

[47].Roma, P.C., Hursh, S.R., Hienz, R.D., Brinson, Z.S., Gasior, E.D., Brady, J.V. (2012). Interactive effects of autonomous operations and circadian factors on crew performance, behavior, and stress physiology. Paper # IAC-12-A1.1.11. International Astronautical Federation. Proceedings, 63th International Astronautical Congress, Naples, Italy, October 1-5, 2012.
[48].Gushin, V., Shved, D., Ehmann, B., Balazss, L., Komarevtsev, S. (2012). Crew-MC interactions during communication delay in Mars-500. Paper # IAC-12-A1.1.2. International Astronautical Federation. Proceedings, 63th International Astronautical Congress, Naples, Italy, October 1-5, 2012.
[49].Sandal, G.M. (2012). "Groupthink" on a mission to Mars: Results from a 520 days space simulation study. Paper # IAC-12-A1.1.3. International Astronautical Federation. Proceedings, 63th International Astronautical Congress, Naples, Italy, October 1-5, 2012.
[50].Van Baarsen, B., Ferlazzo, F., Ferravante, D., Smit, J., van der Pligt, J., van Duijn, M, (2012). The effects of extreme isolation on loneliness and cognitive control processes: Analyses of the Lodgead data obtained during the Mars105 and the Mars520 studies. Paper # IAC-12-A1.1.4. International Astronautical Federation. Proceedings, 63th International Astronautical Congress, Naples, Italy, October 1-5, 2012.
[51].Basner, M., Dinges, D., Mollicone, D., Savelev, I., Ecker, A., Di Antonio, A., Jones, C., Hyder, E., Kan, K., Morukov, B., Sutton, J. (2012). Behaviour, performance and psychosocial issues in space. Paper # IAC-12-A1.1.5. International Astronautical Federation. Proceedings, 63th International Astronautical Congress, Naples, Italy, October 1-5, 2012.
[52].Tafforin, C. (2012). The Mars-500 crew in daily life activities: Ethological study. Paper # IAC-12-A1.1.6. International Astronautical Federation. Proceedings, 63th International Astronautical Congress, Naples, Italy, October 1-5, 2012.
[53].Kanas, N., Salnitskiy, V., Gushin, V., Weiss, D.S., Grund, E.M., Flynn, C., Kozerenko, O., Sled, A., Marmar, C.R. (2001). Asthenia – Does it exist in space? Psychosom. Med., 63, 874-880.
[54].Palinkas, L.A. (1991). Group adaptation and individual adjustment in Antarctica: A summary of recent research. In: A.A. Harrison, Y.A. Clearwater, C.P., McKay, eds. From Antarctica to Outer Space. New York: Springer-Verlag.
[55].Ihle, E.C., Ritsher, J.B., Kanas, N. (2006). Positive psychological outcomes of space flight: An empirical study. Aviat. Space, Environ. Med., 77, 93-101.
[56].Suedfeld, P., Legkaia, K., Brcic, J. (2010). Changes in the hierarchy of value references associated with flying in space. J. Personality, 78(5), 1-25.
[57].Kanas, N. (2011). From Earth's orbit to the outer planets and beyond: Psychological issues in space. Acta Astronaut., 68, 576-581.

Chapter 6.

[1] N. Berend et al. Feasibility Assessment of Rapid Earth–Mars Transfers Using High-Power Electric Propulsion, Journal of Spacecraft and Rockets, Vol. 51, No. 3, pp. 946-957, May–June 2014.

[2] S. K. Borowski, D. R. McCurdy and T. W. Packard, "7-Launch" NTR Space Transportation System for NASA's Mars Design Reference Architecture (DRA) 5.0, 45th Joint Propulsion Conference of the American Institute of Aeronautics and Astronautics, August 2-5, 2009.

[3] R.D. Braun and R.M. Manning, Mars Entry, Descent and Landing Challenges, Journal of Spacecraft and Rockets, Vol. 44 (2), 310–323, Mar-Apr, 2007.

[4] J.R. Brothy, R. Gershman, N. Strange, D. Landau, R. G. Merril, T. Kerslake, 300-kW Solar Electric Propulsion System Configuration for Human Exploration of Near-Earth Asteroids, AIAA, 2010.

[5] J.A. Christian, K. Manyapu, G.W. Wells, J.M. Lafleur, A.M. Verges and R.D Braun, Sizing of an Entry, Descent, and Landing System for Human Mars Exploration, Journal of Spacecraft and Rockets, Vol. 45, No. 1, pp. 130–141, Jan-Feb, 2008.

[6] C. Cooper, W. Hofstetter, J. A. Hoffman, E.F. Crawley, Assessment of architectural options for surface power generation and energy storage on human Mars missions, Acta Astronautica, vol. 7-8, 1106-112, 2010.

[7] M. Dudley-Rowley, S. Whitney, S. Bishop, B. Caldwell and P.D. Nolan: Crew Size, Composition, and Time: Implications for Habitat and Workplace Design in Extreme Environments, SAE 2001-01-2139, 31st International Conference on Environmental Systems, Orlando, FL, July 2001.

[8] L. Bessone and D. Vennemann, CDF Study Report, Human Missions to Mars, Overall Architecture Assessment, CDF-20(A), February 2004.

[9] F. Ferraro, R. D'Elia, S. Paternostro, A. Simonetti, C. Bruno, Analysis of a Manned Mars Mission with Nuclear Electric Propulsion (NEP) System, IAC Paper IAC-09.C4.7.-C3.5.2, Presented at the 60th IAC, Daejeon, Korea, 12–16 October 2009.

[10] L.E. George and L.D. Kos, "Interplanetary mission design handbook: Earth to Mars mission opportunities and Mars to Earth return opportunities 2009-2024", NASA-TM-1998-208533, 1998.

[10] J. Holladay, C. Crumbly, B. Hampton and T. Monk, A Comparison of Future Space Launch System (SLS) Exploration Technologies: In-Space Stages, proceedings of the IEEE Aerospace Conference, paper 2249, Big Sky, Montana (USA), 1-8 March 2014.

[11] N. Kanas et al, Review, Psychology and culture during long-duration space missions, Acta Astronautica, vol. 64, 659–677, 2009.

[12] D. F. Landau and J. M. Longuski, Trajectories for Human Missions to Mars, Part 1: Impulsive Transfers, Journal of Spacecraft and Rockets, Vol. 43, No. 5, September–October 2006.

[13] D.F. Landau and N.J. Strange, "Human Exploration of Near-Earth Asteroids via Solar Electric Propulsion", AAS 11-102, 21st AAS/AIAA Space Flight Mechanics Meeting, New Orleans, LA, February 13-17, 2011.

[14] R. Lenard, Nuclear safety, legal aspects and policy recommendations for space nuclear power and propulsion systems, Acta Astronautica, 59(1), 398-412, 2006.

[15] S. Matousek, B. Sergeyevsky, To Mars and back: 2002-2020 ballistic trajectory data for the mission architect, AIAA98-3736, 1998.

[16] T.A. May and D.S. Creech, NASA's Space Launch System (SLS) Program: Mars Program Utilization, NASA MSC report no 20120015287, 2012.

[17] S. J. Hoffman and D. I. Kaplan, eds., Human Exploration of Mars: The Reference Mission of the NASA Mars Exploration Study Team - NASA SP-6107 - NASA Johnson Space Center, July 1997.

[18] S.J. Hoffman, ed., The Mars Surface Reference Mission: A Description of Human and Robotic Surface Activities, NASA/TP-2001-209371, 2001.

[19] G. Drake ed., Mars Architecture Steering Group, 1st Addendum of the Human Exploration of Mars, Design Reference Architecture 5.0, NASA Johnson Space Center, 2009.

[20] G. Drake ed., Mars Architecture Steering Group, 2nd Addendum of the Human Exploration of Mars, Design Reference Architecture 5.0, NASA Johnson Space Center, 2014.

[21] M. Noca, J.E. Polk and R. Lenard, Nuclear electric propulsion for the exploration of the outer planets, Innovative Approaches to Outer Planetary Exploration 2001–2020, 2001.

[22] J.M. Salotti, Revised Scenario for Human Missions to Mars, Acta Astronautica, vol. 81, p. 273-287, 2012.

[23] J.-M. Salotti and E. Suhir, Manned missions to Mars: Minimizing risks of failure, Acta Astronautica, vol. 93, 148–161, 2014.

[24] J.M. Salotti, R. Heidmann and E. Suhir, Crew Size Impact on the Design, Risks and Cost of a Human Mission to Mars, Proceedings of the IEEE Aerospace Conference, Big Sky, Montana (USA), pp. 1-9, March 2014.

[25] T.D. Schmidt, W. Seboldt and M. Auweter-Kurtz, Flexible piloted Mars mission using continuous electric propulsion, Journal of Spacecraft and Rockets, vol. 43 (6), 2006.

[26] B. St Germain, J.R. Olds, J. Bradford, A.C. Charania, D. DePasquale, M. Schaffer, J. Wallace, Uilizing Lunar Architecture Transportation Elements for Mars Exploration, proc. of the AIAA Space 2007 Conference, Long Beach, California, 18-20 september 2007.

[27] M. Stamatelatos, H. Dezfuli, Probabilistic risk assessment procedures, guide for NASA managers and practitioners, NASA Center for AeroSpace Information, NASA/SP-2011-3421, 2011.

[28] B. Steinfeldt, J. Theisinger, A. Korzun, I. Clark, M. Grant, and R. Braun, High Mass Mars Entry, Descent, and Landing, Architecture Assessment, Proc. of the AIAA Space 2009 Conference and Exposition, AIAA 2009-6684, Pasadena, CA, 14-17 September 2009.

[29] N. Strange, R. Merrill, D. Landau, B. Drake, J. Brophy, and R. Hofer, "Human Missions to Phobos and Deimos Using Combined Chemical and Solar Electric Propulsion", AIAA-2011-5663, 47th AIAA/ASME/SAE/ASEE Joint Propulsion Conference and Exhibit, San Diego, California, July 31-3, 2011.

[30] S. White, B. Spence, T. Trautt, P. Cronin, ULTRAFLEX-175 on Space Technology 8 (ST8) – Validating the Next Generation in Lightweight Solar Arrays, proceedings of NASA Science and Technology Conference, University of Maryland, June 19-21, 2007.
[31] P. D. Wooster, R. D. Braun, J. Ahn, and Z. R. Putnam, Mission design options for human Mars missions, The International Journal of Mars Science and Exploration, vol. 3, 12-28, 2007.
[32] H. Wright, et al. "Mars Aerocapture Systems Study", NASA TM 2006-214522, Nov. 2006.
[33] R. Zubrin and D. A. Baker, Mars Direct: Humans to the Red Planet by 1999, proceedings of the 41st Congress of the International Astronautical Federation, 1990.
[34] R. Zubrin and D. Weaver, Practical Methods for Near-Term Piloted Mars Missions, AIAA 93-2089 AIAA/SAE 29th Joint Propulsion Conference, Monterey CA, 1993.

Chapter 7.

[1]LIQUIFER Systems Group et.al.; "SHEE Self-Deployable Habitat for Extreme Environments: Design Requirements", 2013-2014, p. 11
[2] LIQUIFER Systems Group et.al.; "SHEE Self-Deployable Habitat for Extreme Environments: Design Requirements", 2013-2014, p. 11
[3] Wiley J. Larson, W.J.; Pranke, L.K.; (October 26, 1999) Human Spaceflight: Mission Analysis and Design, McGraw-Hill Companies; 1 edition, October 26, 1999, p. 590.
[4] lmhof, B., "The Socio-Psychological Impact of Architectural Spaces in Long-Duration Missions," SAE Technical Paper 2003-01-2537, 2003, doi:10.4271/2003-01-2537, p. 9
[5] Adams, Constance M. (1998 July). Four Legs in the Morning: Issues in Crew-Quarter Design for Long-Duration Space Facilities (SAE 981794). 28th International Conference on Environmental Systems (ICES), Danvers, Massachusetts, USA, 13-16 July 1998. Warrendale, Pennsylvania, USA: Society of Automotive Engineers, p. 8
[6] LIQUIFER Systems Group et.al.; "SHEE Self-Deployable Habitat for Extreme Environments: Design Requirements", 2013-2014, p. 26
[7] C. Cooper, W. Hofstetter, J. A. Hoffman, E.F. Crawley, Assessment of architectural options for surface power generation and energy storage on human Mars missions, Acta Astronautica, vol. 7-8, 1106-112, 2010.
[8] lmhof, B., "The Socio-Psychological Impact of Architectural Spaces in Long-Duration Missions," SAE Technical Paper 2003-01-2537, 2003, doi:10.4271/2003-01-2537.
[9]LIQUIFER Systems Group et.al.; "SHEE Self-Deployable Habitat for Extreme Environments: Design Requirements", 2013-2014
[10] LIQUIFER Systems Group, "RAMA - Rover for advanced mission applications ESA Surface architecture studies", Doc. No.: LSG-CCN-TN100, 22.07.08
[11] Kriss J. Kennedy, "Advanced Habitation Efforts at JSC-CTSD", *Space Architect* NASA Johnson Space Center, Houston, Texas 77058

[12] Mohanty, S.; Jørgensen, J.; Nyström, M., Ph.D.; Psychological Factors Associated with Habitat Design for Planetary Mission Simulators, Space 2006 AIAA 2006-7345 19 - 21 September 2006, San Jose, California
[13] NASA, "Deep Space Habitat" http://nasa.gov/exploration/technology/deep_space_habitat/
[14] THESEUS, towards Human Exploration of Space a European Strategy - Roadmap, EU-FP 7/2007-2013, grant agreement no. 242482, March 2012, http://www.theseus-eu.org
[15] ESA AO6341-SOW for "Scenario Studies for Human Spaceflight and Exploration" HSF-EA/STU/SOW/BH/2009-18100, Iss.1 Rev.1, 21 December 2009, Sect. 1.4, pp.7-9.
[16] LIQUIFER Systems Group, "RAMA, Concept Study for a Pressurized Rover for Moon and Mars Exploration," Contract No. 20870/07/NL/GM for Thales Alenia Space and ESA CCN to "Analysis of Surface Architecture for European Space Exploration - Element Design," Vienna, November 2008
[17] Cohen, Marc, (1998), Space Habitat Design Integration Issues, 98 1800
[18] NASA/JPL, Mars, 24.4.2013, http://science.nationalgeographic.com/science/space /so-lar-system/mars-article/
[19] Astrium, et al, "European Mars Missions Architecture Study," Final Report to ESA, 2002.
[20] Adams, Constance M. (1998 July). Four Legs in the Morning: Issues in Crew-Quarter Design for Long-Duration Space Facilities (SAE 981794). 28th International Conference on Environmental Systems (ICES), Danvers, Massachusetts, USA, 13-16 July 1998. Warrendale, Pennsylvania, USA: Society of Automotive Engineers
[21] Wiley J. Larson, W.J.; Pranke, L.K.; (October 26, 1999) Human Spaceflight: Mission Analysis and Design, McGraw-Hill Companies; 1 edition, October 26, 1999
[22] Manned mission to Mars, http://en.wikipedia.org/wiki/Manned_mission_to_Mars
[23] Anderson, B.; The Greenhouse Effect - Living with Plants in Other Worlds, http://www.lpi.usra.edu/publications/newsletters/lpib/lpib85/plants.html
[24] Harrison, A.; Clearwater Y.; McKay, C.P.; From Antarctica to Outer Space - Life in Isolation and Confinement, Springer Verlag, 1991
[25] HUMEX, Study no the Survivability and Adaptation of Humans to Long-Duration Interplanetary and Planetary Environments, ESTEC/Cntract No. 14056/99/NL/PA
[26] Kring, J.; Human Consideration in the Design of Closed Environmental Systems, Life Support and Biosphere Science, vol. 7, no. 1, 2000
[27] Adams, C.; McCurdy, M.R.; Habitability as a Tier One Criterion in Advanced Space Vehicle Design, Part One: Habitability (SAE 1999-01-2137). 29th International Conference on Environmental Systems (ICES), Danvers, Massachusetts, USA, 24-26 September 1996. Reston, Virginia, USA: American Institute of Aeronautics and Astronautics.
[28] Connors, M.; Harrison A,; Living Aloft: Human Requirements for Extended Spaceflights, Government Printing Office, 1985, http://history.nasa.gov/SP-483/ch1-4.htm

[29] Imhof, B.; Schartner, H.; Mars Surface Habitats: Architectural Designs and Concepts for Planetary Outposts (SAE 2001-01-2174). 31st International Conference on Environmental Systems (ICES), Orlando, Florida, USA, 9-12 July 2001

[30] Man-System Integration Standard, Vol. 1, NASA STD 3000, Rev.B (1995), NASA-Johnson Space Center, Houston, TX

Chapter 8.

[1] K. A. Binsted and J. B. Hunter , HI SEAS (Hawaii Space Exploration Analog and Simulation, hi-seas.org) as an opportunity for long-duration instrument/protocol testing and verification, Proceedings of the 2nd conference on Analog Sites for Mars Missions II: past, present and future missions to Mars, Washington, DC - August 5th to 7th 2013.

[2] R. D. Carpenter, T.F. Lang, S. A. Bloomfield, J. J. Bloomberg S. Judex, J. H. Keyak, R. J. Midura, P. Divieti Pajevic, J. M. Spatz, Effects of Long-Duration Spaceflight, Microgravity, and Radiation on the Neuromuscular, Sensorimotor, and Skeletal Systems, Journal of Cosmology, Vol 12, 3778-3780, 2010.

[3] O. Doule, B. Imhof, W. Hohender, S. Ransom, A. Aablo, J. Nelson, V. Šálený, M. Ilzkovitz, J. Gancet, B. Gardette, V. Taillebot and P. Weiss, Self-deployable Habitat for Extreme Environments - Innovative Architecture Test-bed for Terrestrial and Space Applications, 64th International Astronautical Congress, Beijing, China, 2013.

[4] D. Eppler and J. Bleacher, NASA's 2010 Desert Research and Technology Studies Mission Objectives and Results, Acta Astronautica, Vol. 90, Issue 2, 181-416, October 2013.

[5] G. Groemer et al, The MARS2013 Mars Analog Mission, Astrobiology, 14(5), 360-376, May 2014.

[6] A. J. Hanford, Advanced Life Support Research and Technology Development Metric, Rapport NASA n°NASA/CR-2006-213694, February 2006.

[7] C. Lasseur, J. Brunet, H. de Weever, M. Dixon, G. Dussap, F. Godia , MELiSSA: The European Project of Closed Life Support System, Gravitational and Space Biology 23(2) August 2010.

[8] Mars Desert Research Station reports, available at http://mdrs.marssociety.org/, accessed December 2014.

[9] Marshall Space Flight Center, "Designing for Human Presence in Space: Environment Control and Life Support System", NASA RP-1324, September 1994.

[10] NASA analog missions, Paving the way for space exploration, NASA Langley Research Center, report NP-2011-06-395-LaRC, 2011.

[11] R. Rafkin, A positive radiative-dynamic feedback mechanism for the maintenance and growth of Martian dust storms, Journal of Geophysical Research, vol. 114, E01009, 2009.

[12] K. R. Sridhar, J. E. Finn, and M. H. Kliss, In-Situ Resource Utilization Technologies for Mars Life Support Systems, Advances in Space Research, vol. 25, 2, pp. 249-255, 2000.

[13] D. Urbina and R. Charles, Symposium keynote: Enduring the isolation of interplanetary travel: A personal account of the Mars500 mission. Acta Astronautica, 93, 374-383, 2014.

[14] P. Weiss, B. Gardette, B. Chirie, J. Collina-Girard, H.G. Delauze, Simulation and preparation of surface EVA in reduced gravity at the Marseilles Bay subsea analogue sites, Planetary and Space Sciences, 74(1), 121–134, December 2012.

[15] P.O. Wieland, Designing For Human Presence in Space: An Introduction to Environmental Control and Life Support Systems (ECLSS), Appendix I, Update—Historical ECLSS for U.S. and U.S.S.R./Russian Space Habitats, rapport MSFC/NASA n° NASA/TM—2005–214007, 2005.

[16] R. Zubrin, In-Situ Propellant Production: The Key Technology Required for the Realization of a Coherent and Cost Effective Space Exploration Initiative, Paper No. IAA 91-668, 42nd Congress of the Intenational Astronautical Federation, Montreal, Canada, 1991.

Chapter 9.

[1] D. Beaty, M. Grady, L. May and B. Gardini, Preliminary Planning for an International Mars Sample Return Mission, Report of the International Mars Architecture for the Return of Samples (iMARS) Working Group, June 1, 2008.

[2] C. Culbert, O. Mongrard, N. Satoh, K. Goodliff, C. Seaman, P. Troutman and E. Martin, ISECG Mission Scenarios and their Role in Informing Next Steps for Human Mission Exploration Beyond Low Earth Orbit, proceedings of the 62nd Int. Astron. Congress, IAC-11-D3.1.2, October 2011.

[3] B. Hufenbach, K. C. Laurini, J.-C. Piedboeuf, B. Schade, K. Matsumoto, F. Spiero, A. Lorenzoni, The Global Exploration Roadmap, proceedings of the 62nd Int. Astr. Congress, IAC-11.B3.1.8, Cape Town, October 2011.

[4] International Space Exploration Coordinating Group, The Global Exploration Roadmap, ISECG Technical Report, August 2013.

[5] D.J.P. Moura and the iMARS Team, The Road to an International Architecture for Mars Sample Return: The IMARS Team View, proceedings of the Int. Astron. Congress, IAC-08-A3.1.3, October 2008.

[6] J.M. Salotti and R. Heidmann, Roadmap to a Human Mars Mission, Acta Astronautica, Volume 104, Issue 2, p. 558-564, 2014.

[7] J.M. Salotti, R. Heidmann and E. Suhir, Crew Size Impact on the Design, Risks and Cost of a Human Mission to Mars, Proceedings of the IEEE Aerospace Conference, Big Sky, Montana (USA), pp. 1-9, March 2014.

APPENDIX A - LIST OF PARTICIPANTS TO THE STUDY GROUP

Co-chairmen: Giancarlo Genta and Alain Dupas
Secretary: Jean-Marc Salotti

Chapter editors:

Chapter 1: Richard Heidmann
Chapter 2: Alain Dupas
Chapter 3: Julien Alexandre Lamamy
Chapter 4: Giancarlo Genta
Chapter 5: Nick Kanas & Susan McKenna-Lawlor
Chapter 6: Andreas Rittweger & Jean-Marc Salotti
Chapter 7: Maria Antonietta Perino
Chapter 8: Jean-Marc Salotti
Chapter 9: Alain Dupas & Jean-Marc Salotti
Chapter 10: Giancarlo Genta

Members: (in alphabetic order):
Mauro Augelli (CNES, France)
Andrew Aldrin (ULA, USA)
Nicolas Bérend (ONERA, France)
Giovanni Bignami (COSPAR)
Christophe Bonnal (CNES, France)
Claudio Bruno (University of Rome, Italy)
John B. Charles (NASA, USA)
Lin Chen (Lockheed Martin, USA)
Elisa Cliquet Moreno (CNES, France)
Gabriel G. De la Torre (Univ. of Cadiz, Spain)
Nadeem Ghafoor (MDA, Canada)
Alexey Grushevski (Russia)
Richard. Heidmann (Planet Mars Ass., France)
Bernhard Hufenbach (ESA, Europe)
Les Johnson (NASA, USA)
Nick Kanas (U. of California, S. Francisco, USA)
David Kendall (Canadian Space Agency, Canada)
Andrey A. Kuricyn (Russia)
Julien-Alexandre Lamamy (Orb. Sc. Corp. USA)
John Logsdon (George Washington Univ., USA)
Kohtaro Matsumoto (JAXA, Japan)
Chris McKay (NASA, USA)
Susan McKenna-Lawlor (Space Tech.Ireland Ltd. Ireland)
Ernst Messerschmid (Stuttgart U., Germany)
Natalia Nikolaeva (Russia)
Gian Gabriele Ori (Univ. G. d'Annunzio, Italy)
Maria Antonietta Perino (TASI, Italy)
Viacheslav Petukhov (Russia)
Yury Razoumny (Russia)

Pascal Renten (University of Stuttgart, Germany)
Stephen C. Ringler (SGAC, USA)
Andreas Rittweger (Astrium, France)
Christian Sallaberger (MDA, Canada)
Klaus Schilling (Univ. of Würzburg, Germany)
Alexander Semenkin (Russia)
Carol R. Stoker (NASA, USA)
Liu Wei (China)
Luo Wencheng (China)
Alan Wilhite (Georgia Tech, USA)
Cao Xiaohui (China)
Lu Yu (China)

APPENDIX B - A LIST OF RELEVANT OPTIONS FOR THE MISSION ARCHITECTURE

#	CHOICE	OPTIONS	MAIN CONSIDERATIONS
1	Trajectories scheme	Opposition Conjunction Hohmann Conjunction Free-Return Dash-Flyby	Mission reliability Mission recurrent cost (IMLEO) Safety
2	Overall redundancy	No overall redundancy 2 identical missions in parallel	Mission reliability Mission recurrent cost (IMLEO) Safety
3	Crew size	2 + 2 4 6 Other	Mission efficiency Mission reliability Mission recurrent cost (IMLEO)
4	Launcher category	Medium lift Heavy lift	Mission reliability (launches, RV, launch period) Mission recurrent cost
5	Number of Earth-Mars transfers per mission	One (all-up) Two (split) More than two	Mission reliability (launches, RV, launch period) Mission recurrent cost Safety (automatic infrastructure prepositioning)
6	Crew launch	Separately In one of the main launches	Safety Launcher(s) development and production costs
7	Parking orbit	LEO Nuclear safe orbit (>800 km?) Highly elliptical Earth-Moon Lagrange point	
8	Earth-Mars transfer propulsion	Chemical (H_2/O_2) Nuclear Thermal (NTP) Electrical (NEP or SEP) Mixed Electrical/Chemical	Safety Mission reliability Mission recurrent cost (IMLEO) Development cost Political acceptance (nuclear devices)
9	Earth-Mars transfer gravity environment	No artificial gravity Complete spacecraft spinning Centrifuge with tether	Safety Crew comfort
10	Mars orbit insertion	w/o (direct descente) Propulsive insertion Aerocapture	Mission recurrent cost (IMLEO) Safety Flexibility

Appendix B. A list of relevant options for the mission architecture

11	Descent vehicle	Interplanetary Hab Special S/C prepositioned - transferred with Hab	Mission recurrent cost (IMLEO) Safety Development cost
12	EDL aerodynamic formula	Lifting body Capsule with aeroshell	Development cost Delay risk
13	EDL aerodynamic shell	Separate aeroc. & EDL shells Unique aerodynamic shell	Mission recurrent cost (IMLEO)
14	Aerocapture & EDL shells technology	Fixed Deployable Inflatable	Development cost Delay risk
15	EDL Auxiliary aerodynamic devices	No Parachute only Parachute and Ballute	Development cost Delay risk Mission recurrent cost
16	EDL propulsion use	Final (subsonic) only Supersonic and final	Safety Delay risk
17	EDL propulsion type	Cryo H_2/O_2 CH_4/O_2 Storables	Development cost Mission recurrent cost (IMLEO)
18	Multiple missions strategy	Separate remote landing sites Separate sites within rover range One unique (growing) site	Safety Mission reliability Scientific productivity
19	Crew surface Hab	Prepositioned Hab Descent S/C	Safety Development cost
20	Main surface energy source	Nuclear prepositioned Nuclear man-deployed Solar predeployed Solar man-deployed	Development cost Delay risk Political acceptance (nuclear devices)
21	Safety supplies cache	No Consumables only Consumables & spare parts	Safety
22	Hab water management	No recycling Recycling to ISS state of art Advanced recycling ratio	Safety Development cost Delay risk Mission recurrent cost (IMLEO)
23	Hab air management / O_2	Bioregeneration CO_2 physicochemical decomp. O_2 production from Mars atm.	Safety Development cost Mission recurrent cost(IMLEO)
24	Hab air management / CO_2 removal	Bioregeneration LiH capsules CO_2 physicochemical decomp.	Safety Development cost Mission recurrent cost(IMLEO)
25	Air pressure and composition	1 atm Earth-like 0.5 atm O_2/N_2 0.35 atm O_2/N_2 (Skylab heritage) Pure O_2	Mission recurrent cost(IMLEO)

26	Food origin	Fully Earth produced inventory Partly greenhouse produced	Development cost Mission recurrent cost(IMLEO)
27	Food conservation	Lyophilized only Lyophilized & congealed	Safety Crew comfort
28	Thermal management	Tolerant to failure Not tolerant	Safety
29	Waste management	Dump Uses for shielding Edible waste in greenhouse	Development cost Mission recurrent cost
30	Health management / Surgery level	Low Advanced	Safety Mission recurrent cost(IMLEO)
31	Health management / Medicines supplement	Most probable needs Less probable but critical needs	Safety Mission recurrent cost(IMLEO)
32	GCR mitigation during transfers (shelter provided against SPE anyway)	No Partial H_20 / $(CH_2)_n$ shielding	Crew long-term health
33	Greenhouse	No Experimental & psychological Full ECLSS	Development cost Mission recurrent cost (IMLEO) Delay risk
34	Earth-Mars telecommunications	Radio (band) Laser	Development cost Safety Mission reliability
35	Surface mobility	Unpressurized rovers only Unpressurized & pressurized rovers	Development cost Mission recurrent cost (IMLEO and Hardware) Safety Scientific productivity
36	Pressurized rover size	2 crew (nominal) 3 crew (nominal)	Scientific productivity Safety Mission recurrent cost (IMLEO)
37	Pressurized rover range	Safe quad return distance 100 km >100 km	Scientific productivity Safety
38	Pressurized rover motorization	Electrical with Batteries Electrical with Fuel Cells Internal Combustion Engines	Development cost Delay risk Scientific productivity (through reliability)
39	Hab scientific equipment complement	In-depth analyses on Earth only Minimal complement allowing on-site sorties plan adaptation Full complement for in-depth geological and biological analyses	Scientific productivity Mission recurrent cost (IMLEO, hardware)

40	Drilling capacity	Handheld Light drilling station (few meters) Heavy motorized station (rover)	Scientific productivity Mission recurrent cost (IMLEO, hardware)
41	In Situ Propellant Production	None Limited to O_2 $O_2 + CH_4$ (or CH_3OH…)	Mission recurrent cost (IMLEO, hardware) Safety Delay risk Development cost
42	Mars ground ice utilization	None For water For water, O_2, H_2	Mission recurrent cost (IMLEO, hardware) Safety
43	Earth Return mode	Direct Mars orbit RV	Safety Mission recurrent cost (IMLEO, hardware)
44	Ascent propulsion / Propergol	O_2/CH_4 Storables	Safety Mission recurrent cost (IMLEO, hardware)
45	Ascent propulsion / Feed mode	Pressure fed Turbopumps	Safety Mission recurrent cost (IMLEO, hardware)
46	Ascent vehicle	Prepositioned Same as descent vehicle	Safety
47	Mars-Earth transfer propulsion	Chemical (CH_4/O_2) Nuclear Thermal Electrical Mixed Electrical/Chemical	Safety Mission recurrent cost (IMLEO) Development cost Political acceptance (nuclear devices)
48	Earth Return	Direct, sea-landing Direct, ground-landing Earth orbit Earth-Moon Lagrange point	Safety Flexibility
49	Program management	One global agency Several coordinating agencies	Development cost Over cost risk Delay risk Geostrategic considerations
50	International industrial organization	One prime with subcontractors Several "separate" industrial primes	Development cost Over cost risk Delay risk Geostrategic considerations
51	System-level testing environment	Earth / Earth orbit / The Moon Deep space (Lagrange, asteroids) Mars orbit	Development cost Over cost risk Delay risk

APPENDIX C - The different propulsion systems, how they work, what is their development status and associated challenges

C.1 Cryogenic propulsion

Cryogenic propulsion is one of the varieties of chemical propulsion. The thermal energy generated by combustion of hydrogen and oxygen is converted into kinetic energy by expansion in a nozzle. The combustion of hydrogen and oxygen is particularly effective for propulsion because it allows a very high specific impulse (around 450s) compared to other propellant types, thanks to its high combustion temperature (around 3600K) and the lightness of the combustion products (mainly water vapor). However, hydrogen and oxygen are in the gaseous phase at room temperature while it is absolutely necessary that they remain in liquid form to have tanks of reasonable sizes, so they must be stored at extremely low temperatures: 21K for hydrogen and 90K for oxygen.

This technology is perfectly mastered today for launchers propulsion (e.g., for Ariane 5). It was also already used for the Apollo missions (J2 engine) for the trans-lunar injection. The challenge of using cryogenic propulsion for a human mars mission lies in two aspects:

- Mastering the evaporation of the propellants in orbit for the duration of the mission: today we are able to store liquid hydrogen and oxygen in space for a few weeks, but for a human mars mission several months of storage in space may be needed (if it is foreseen to use it for Mars orbit insertion or descent) and even more (>1 ½ years) if this propulsion is also considered for the return trajectory. There are however ongoing works at agency level (NASA, CNES) to improve tank insulation and to make active cooling of the tank possible and efficient. The associated roadmap is clear and underway.
- Keeping a low IMLEO for the mission: indeed, if used alone for all the main propelled phases of the mission (trans-Mars injection, mars orbit insertion, trans-Earth injection), total IMLEO for a crew of six will not be far from 1000 tons. Thus, to be interesting in terms of IMLEO (e.g., less than ISS mass), cryogenic propulsion absolutely has to be combined with ISPP for the return and aerocapture for Mars orbit insertion.

C.2 Nuclear thermal propulsion

From outside, a nuclear thermal propulsion stage looks like a chemical propulsion stage with a single tank: the tank feeds an engine with a shape that resembles that of a conventional engine but with a very large chamber. The "chamber" contains a fission reactor releasing a few hundred megawatts to heat up the liquid hydrogen stored in tanks. As the engine nozzle heats up, the cooled hydrogen is pumped through the coolant passages in the nozzle walls, before being pumped into the ignition chamber.

Hydrogen is heated to about 2300K in the reactor before being accelerated in the nozzle. The ejected gas is cooler than the one ejected from a cryogenic engine but it is much lighter:

its specific impulse (900s) is roughly twice the one of cryogenic engine. However, the mass of the system is heavier, hydrogen tanks have a poorer structural index than oxygen tanks, and the reactor itself is very heavy compared to a combustion chamber. Moreover, a shield protecting the payload and the crew from neutron and gamma rays has to be added to the system, this shield can be very heavy.

NTP has been extensively tested in the US (ROVER and NERVA programs, both terminated in 1972) and in USSR between 1960 and the Early 1970s. Some engines have operated up to one hour with a thrust of 330kN. In 1972, technology was almost ready to fly but the program was stopped due to the absence of a human Mars program. NTP is considered to be one of the key technologies for a Mars mission by NASA and research continues within NASA and the Department of Energy (DOE) laboratories. The challenges associated with the use of NTP are the following:

- NTP is a single confinement barrier nuclear system: the only barrier separating the uranium and fission products from the environment is the fuel cladding. The cladding technology is highly challenging: it must be resistant to extremely high temperatures in hydrogen atmosphere, which is known to weaken the material. Moreover, the development of NTP is difficult due to ground testing problems. Testing a NTP device, as it was done in the past, was like testing a nuclear power plant with the primary circuit open. Today, all nuclear systems must have at least three confinement barriers (fuel cladding, primary circuit envelope, and confinement building of the reactor). Having an environmentally safe but still representative test of a nuclear thermal propulsion system is thus a complex problem. In the USA, the plan is to test the system in a borehole in the desert [5], but a non-nuclear sub-scale demonstration of the limited propagation of undesirable elements is needed and might reveal the unacceptability of such a test stand. Even if this test layout were chosen, it would probably be difficult to test the engine with a representative duration in one shot due to pressure increase in the hole during the test. If this test procedure were not considered environmentally acceptable, a brand new test facility in closed loop would be necessary this facility would be costly while still not alleviating the difficulties of producing the appropriate representative vacuum or low pressure condtions at the nozzle outlet. As a consequence, testing of nuclear thermal propulsion is difficult.
- The necessity of keeping the nuclear system at a sufficiently high orbit. When discussing space nuclear propulsion in general, some principles have to be respected, notably the 47-68 UN-COPUOS (Committee on the Peaceful Uses of Outer Space) resolution which states that reactors may be operated in 'sufficiently high orbits' with the following definition "The sufficiently high orbit is one in which the orbital lifetime is long enough to allow for a sufficient decay of the fission products to approximately the activity of the actinides. The sufficiently high orbit must be such that the risks to existing and future outer space missions and of collision with other space objects are kept to a minimum. The necessity for the parts of a destroyed reactor also to attain the required decay time before re-entering the Earth's atmosphere shall be considered in determining the sufficiently high orbit altitude." These systems may also be operated in low-Earth orbits if they are stored in sufficiently high orbits after the operational part of their mission. Compared to other

reactors, nuclear thermal cores have an extremely low burn-up because they operate for only a few hours. Thus the inventory of fission product is very limited and it might be possible to start them in LEO since they have high thrust and should thus be able to move out of Earth's sphere of influence more quickly than low thrust systems. However, there is no consensus yet in the community on this possibility and it might be required to start it anyway from a sufficiently high orbit, not for the consideration of the fission product inventory but also to avoid reentry and collision in case of failure of the system to carry out its mission. This would really penalize the NTP system because it would reduce the advantage it has over cryogenic propulsion in term of ISP, since an extra stage of conventional propulsion (or extra capacity of the launcher, or electric tug) would be needed.

- Public acceptance of the use of nuclear systems in space . Use of nuclear power in space is more likely to be accepted by the general public if there is no alternative solution (like for the radioisotope generators beyond Jupiter) or if the nuclear solution offers a very clear advantage compared to the other options. This is not the case as there is no consensus in the scientific community on the superiority of an NTP based scenario over the ones based on cryogenic propulsion combined with in situ propellant production and aerocapture.
- Moreover, and more specifically for nuclear thermal propulsion: today, the US seems to be the only nation actively working on NTP. In the frame of a mission that would be done in international cooperation, this has to be considered. Other participating countries may lobby in favor of other propulsion systems because they would prefer a propulsion type where they also have expertise, or because they have a clear reluctance in their country on the use of nuclear power if alternative solution exists.
- Launch periods: This issue is similar to the one encountered for cryogenic propulsion. However, with the increased ISP and disregarding aerocapture, the problem is less critical for NTP than for chemical propulsion.

C.3 Nuclear electric propulsion

A NEP system is composed of two main entities: a fission-based power generation system, and an electric propulsion module.

Today probes and satellites are equipped with electric thrusters fed by solar panels capable of providing them a few kilowatts of power. The operation of electric thrusters do not depend on the power source, however. In these systems a neutral gas is ionized and then accelerated either by an electric field acting on the ions (gridded ion thrusters, Hall-effect thrusters), or by a magnetic field acting on the plasma (magnetoplasmadymanic thrusters, VASIMR, etc.). These methods allow for ISPs that are larger by one order of magnitude than those of chemical engines (depending on the gas and technology used, this could range from thousands to tens of thousands. However, the power available today with solar panels only allows for the ionization of a very small gas flow rate, which negatively impacts the system's thrust which is directly proportional to the power. Using today's technology, this leads to thrusts in the range of only a few dozen millinewtons, for example, the PPS 1350 used on Smart-1 produced only 88mN for 1.5kW).

For a human Mars mission, several tons of equipment have to be sent to Mars. In order to make the journey in a comparable time frame to chemical or NTP (roughly 6 month in case of a conjunction class mission), not only would it be necessary to be able to provide 5,000 to 10,000 times more power to reach several hundreds of newtons, but new, bigger engines would have to be qualified to limit the number of engines on the vehicle. To provide the necessary megawatts of electricity, the use of a nuclear power generation system would be a lighter, more practical solution than hypothetical solar panels that would need dozen of thousands of square meters, even with improved efficiency compared to today's panels.

A nuclear power generation system is based on the production of heat by a small fission reactor (some dozens of megawatt of thermal power), this heat is transported by a fluid to a conversion system and the extra heat not converted in electricity is rejected through radiators. A shield, which can be as heavy as the reactor itself, protects the payload and the crew from the neutrons and the gamma rays.

The main challenges associated to a nuclear electric propulsion system for human Mars mission are the following:

- To achieve a specific mass in the range of 10 kg/kWe at 10MWe in order to be able to complete a round trip mission with a conjunction like profile. Trajectory calculation shows that this specific mass is needed to achieve a six-month trip time [10]. This specific mass is already challenging and would probably need more than 10 years of development. Indeed, in order to reach this objective, the core has to operate at a temperature of 1300K (no nuclear fuel is qualified today at this temperature). A conversion system, based on a closed Brayton cycle would match this requirement, but turbine operation during thousands of hours at this operating temperature yet has to be demonstrated.
- A few thousand square meters of radiant panels will be needed: these will have to be lightweight, foldable to fit in the shroud of the launcher, and resistant to potential small meteorite impacts. A ground prototype of the full system will certainly be needed. Such technology and facility developments will of course be very expensive (roughly similar to the cost of a facility for a research reactor). However, development cost could be partially covered by other programs thanks to the numerous application of nuclear electric propulsion. Russia is aiming at a ground demonstration in 2018 of a nuclear power and propulsion system (NPPS) at 1MW.
- NEP is sometimes envisioned as a means to reduce transfer times to Mars compared to chemical or nuclear propulsion. It would indeed be possible to reduce the transfer time to four months if a specific mass of 5kg/kWe could be achieved. This could be done by developing a system using the two phase flow Rankine conversion which allows for a dramatic decrease of radiator size. The main problem with this very attractive option is that the system would need to be tested in space because a two phase flow system operation in zero gravity is not likely to be qualified with ground test only. Further reduction of the specific mass could allow for shorter transfer times but seems too technologically out of range to be considered for first missions to Mars (gaseous cores operating at 2000K or more would be needed, such nuclear fuels do not even exist on ground).
- To qualify very high power thrusters: today there is only a limited number of entities working on really high power thrusters (>100kWe). Candidate thrusters are mainly ion

engines, magnetoplasma dynamic engines and other engines such as VASIMR. Self-field magneto plasma dynamic thrusters already have achieved significant steps at very high power, for example they have been tested in ranges from 200kWe to 1MWe at the Institut für Raumfarhtsysteme in Stuttgart, in Princeton university and in Russia (Keldysh Research Center, Fakel, Energya and Moscow Aviation Institute) where a Lithium Lorentz force accelerator has been tested at 500kWe during a 500 hour experiment. VASIMR engines have been tested at 200kWe but only for roughly one minute. The main difficulty of those ultra-powerful thrusters lies in their endurance: long term thermal management (thousands of hours are requested) is far from being mastered today; no engines have been tested with a representative cooling function. Another big challenge lies in the vacuum chambers: no current vacuum chamber has the capability to allow for representative conditions in term of lifetime and vacuum. Such engines might have to be tested in space. Moreover, today, it is very difficult to give a specific mass figure for a flight engine.

- Use of nuclear power in space: public acceptance of the use of a nuclear reactor in space seems more favorable to NEP compared to NTP. Nuclear electric propulsion is of very high interest for a wide range of mission applications, which could absolutely not be done without on board nuclear power (outer planet mission). As a consequence, it is likely that a smaller less powerful system will be developed first to demonstrate the capability and interest of this type of propulsion and pave the way for the use of it for human Mars missions. Compared to NTP and chemical propulsion, NEP allows for an all-up mission where crew and cargo are sent together with a competitive IMLEO in the range of 300 to 400 tons. This is attractive in term of crew safety, as they have all their provisions, materials, and propellants with them at all times. The availability of return propellant could provide interesting abort options. Such a vehicle would also allow for more flexible launch periods and launch opportunities thanks to the very high ISP of electric thrusters, which also compensates for the need to start the system only on a sufficiently high orbit. Even considering those advantages, public acceptance issues might arise, and chemical propulsion is clearly advantageous from this point of view.

C.4 References

[1] George, L.E., Kos, L.D., "Interplanetary mission design handbook: Earth to Mars mission opportunities and Mars to Earth return opportunities 2009-2024", NASA-TM-1998-208533, 1998

[2] Matousek S., Sergeyevsky. B., "to Mars and back: 2002-2020 ballistic trajectory data for the mission architect", AIAA98-3736

[3] Wooster, P.D., Braun, R.D., Ahn, J. and Putnam, Z.R., "Mission design options for human Mars missions", The International Journal of Mars Science and Exploration, 2007

[4] M. Auweter-Kurtz,C. Bruno,D. Fearn,H. Kurtz,T.J. Lawrence,R.X. Lenard, IAA COMMISSION III SG 2 – NUCLEAR SPACE POWER AND PROPULSION

[5] Howe, S.D., "Ground Testing a Nuclear Thermal Rocket: Design of a Sub-scale Demonstration Experiment", Joint Propulsion Conference, (AIAA-2012-3743)

[6] Lucas, "Dose calculation for nuclear thermal rocket exhaust", Nuclear and Emerging Technologies for Space (n°6803, Lucas, DOE),

[7] Griffin, B. et al., "A comparison of transportation systems for human missions to Mars", AIAA 2004-3834, 40th AIAA/ASME/SAE/ASEE Joint Propulsion Conference and Exhibit, 11-14july 2004, Fort Lauderdale, Florida

[8] 47-68 UN-COPUOS « Principles relevant to the use of nuclear power sources in outer space »

[9] Koppel C. et al., « Disruptive Propulsive Technologies for European Space Missions", EUCASS 2013

[10] Berend, N., "How fast can we go to Mars using high power electric propulsion?" 46th AIAA/ASME /SAE/ASEE Joint Propulsion Conference and Exhibit, Atlanta, AIAA-2012-3889

[11] Clark, J.S., George J.A., Gefert, L.P. "Nuclear Electric Propulsion, A "Better, Safer, Cheaper" Transportation system for Human Exploration of Mars", 11th symposium on Space Nuclear Power and Propulsion Systems, Albuquerque New Mexico, January 9-13,1994, NASA TM 106406

[12] B. Cothran, B. Tillotson and B. Donahue, " Advanced Options for Human exploration of Mars" Boeing Aerospace and electronics, AIAA Space Programs and Technologies Conference, Huntsville, AIAA -90-3822, Sep 25-28,1990

[13] R.J.Cassady, "An MPD thrusters driven cargo ferry for support of the manned Mars Mission", AIAA-88-2896, AIAA/ASME/ SAE / SEE 24th joint propulsion conference, Boston Massachusetts, July 11-13, 1988

[14] Jonathan A. Webb and Brian J. Gross, "Conceptual multi megawatt system based on a tungsten CERMET reactor", Nuclear and Emerging Technologies for Space 2011

[15] Litchford R.J, Harada, N., "Multi MW closed cycle MHD nuclear space power via nonequilibrium He/Xe frozen inert plasma", Nuclear and Emerging Technologies for Space 2011

[16] Smith, B., Knight, T., Anghaie, S., "Multimegawatt NEP with vapour core reactor MHD", STAIF 2002

[17] Woodcock, G., "Evaluation of high performance space nuclear electric generators for electric propulsion application", STAIF 2002

[18] Rovang, R.D., Johnson, G.A., Mills, J.C., "Potassium-Rankine nuclear electric propulsion for Mars cargo missions", STAIF 2002

[19] Longhurst, G.R., Schnitzler, B.G., Parks, B.T., "Multi-Megawatt Power system trade study", STAIF 2002

APPENDIX D - Human Mars Mission Feasibility Index

The concept of Human Mars Mission Feasibility Index (HMMFI) can include technical, human, programmatic, political, and sustainability parameters. Each year it may be updated to take the evolution of the different parameters into account.

As an example, the parameters may be:

1) Technical Parameters, max 10:
- Environment Control and Life Support System (ECLSS)
- Propulsion
- Mission design
- Launcher
- Landing Technology
- Reentry

2) Human Parameters:
- Radiation
- Low gravity
- Psychological
- Physiological

3) Programmatic Parameters
- Political climate for cooperation to a human Mars mission
- Governance of a global cooperation

4) Sustainability Parameters
- Commercial market probability of a human Mars mission
- Government budget affordability

This is just a preliminary suggestion, and a detailed study can lead to identifying a larger number of relevant parameters. Some parameters may be considered irrelevant: for instance, it may be decided that it is possible to go to Mars without any form of artificial gravity, so the parameter about low gravity can be neglected.

Proposal 1: averaging algorithm

Different types of parameters can be suggested. A first possibility is using an index going from 1 to 9, like in the TRL, or 1 to 10. While in Technical Parameters the same definitions as in TRL might be used, it is likely that different definitions must be taken for Human Parameters, Programmatic Parameters, and Sustainability Parameters. It must be noted that even for the TRL different definitions are used in the different industries, or even in a given field by different organizations.

Once the parameters have been decided and a value is assessed for all of them, the simplest thing is to sum them, or to compute an average value. This has however an intrinsic drawback: all parameters are considered equally important.

A better solution may be to perform a weighted average, but to do this the weights have to be stated, something that requires a deep study and, above all, introduces arbitrary evaluation (which parameters are most important?)

Proposal 2: multiplying algorithm

Another suggestion, which has not been deeply discussed in the study group but is shown here as a possible example of an implementation of the index, is that of assuming that all systems proposed to implement a human Mars mission are technically feasible and consider two parameters:

- The development cost, tests and qualification included
- The time needed to achieve that development.

These parameters are thus not primarily technical but essentially financial and political. The feasibility index of a given system (which is already assumed technically feasible) can be defined by the probability that its development and qualification are financed. Intuitively, in first approximation, it is inversely proportional to the costs and to the duration of the development and qualification of that system.

The two parameters are numerically evaluated as

- *C*: the cost in billion dollars required to reach technical feasibility (TRL = 9)
- *T*: number of years for the development and qualification.

Obviously these definitions are completely arbitrary and the evaluation of the value of the parameters has a high degree of subjectiveness. They can be put together issuing an empirical equation like:

$$F_i = \frac{1}{\frac{C_i T_i}{100} + 1} \tag{1}$$

Which yields the feasibility index of the *i*th parameter. The feasibility is thus zero for a very immature technology (infinitely expensive and very long qualification phase) and one for a technology which is ready to use (very cheap and short development and qualification).

The denominator 100 is a completely arbitrary number and is also related to the units used for expressing costs and development time.

The global feasibility index of a mission can be given by the product of all systems feasibility indices:

$$F = \prod F_i \tag{2}$$

Also combining the various indices by multiplying them is completely arbitrary, and is mostly justified by the fact that to have a global index equal to 1 (Mission feasible) it is necessary that all indices are equal to 1 (all components are ready) and that even a single index equal to 0 (a single component unfeasible) makes the global index equal to 0 (mission unfeasible)

Clearly, this approach is much more justified in case of technical parameters than in the case of parameters of other types, but can be adjusted to suit all cases. Maybe in other cases instead of the product of cost and time it is possible to devise a product of other features.

Another point is that the qualification phase of several systems might be carried out in the same mission (for instance a Heavy Mars Sample Return Mission can contribute to the qualification of TMI propulsion stages, aerocapture, EDL systems and ISRU), so that there may be possible savings, which are neglected in this formulation but should be taken into account.

An example including thirteen technological factors is reported in Table D.1. From the table it is clear that this index is strongly nonlinear, so that a case whose feasibility is not very problematic seems to have a very low index. This, however, may be not a big problem if the index is used to compare different solutions or to follow the change in time of the index.

Main systems	Cost $B	Time years	Feasibility. index
Heavy launcher and ground infrastructures	10	5	0,667
Space assembly and staging operations	1	3	0,971
Interplanetary propulsion stages	5	5	0,8
Deep space habitat	5	5	0,8
Aerocapture	5	5	0,8
Entry, descent and landing systems	15	15	0,308
Surface habitat	2	5	0,909
Surface power and energy management	2	5	0,909
Surface mobility, including robotics	1	3	0,971
ISRU, O2 for MAV, surface power included	1	5	0,952
Mars ascent vehicle	3	10	0,769
Earth return vehicle	10	10	0,5
Long duration LSS and human factors	5	10	0,667
Global indicators	Total cost 65 billion $	Min. durat. (if optim.): 15 years	Global feasibility index 0,02

Table D.1. Example of Human Mars Mission Feasibility Index.

The nonlinearity of this definition of the index makes it impossible to compare the index obtained in this way with an index obtained using global parameters (for instance, by assuming a cost of 65 billion dolalrs and a time of 15 years yields an index equal to 0.093, much higher than the mentioned value of 0.02

Conclusion

The two mentioned examples are just a first attempt to define the Human Mars Mission Feasibility Index. If this index is found to be useful, a discussion with all the interested parts must be started and a serious study must be undertaken. This might be a task for a follow-on study of the IAA.

International Academy of Astronautics (IAA)

A Brief Description

Founded:
16 August 1960, Stockholm, Sweden, by Theodore Von Karman. Independent non-governmental organization recognized by the United Nations in 1996.

Aims:
Foster the development of astronautics for peaceful purposes; Recognize individuals who have distinguished themselves in space science or technology; Provide a program through which members may contribute to international endeavors; Promote international cooperation in the advancement of aerospace science.

Structure:
Regular Meeting; Board of Trustees consisting of: President; four Vice-Presidents and twenty-eight Trustees, seven from each Section: Basic Sciences, Engineering Sciences, Life Sciences and Social Sciences. Current President: Dr. Peter Jankowitsch, Austria, Past-President: Dr. Madhavan G. Nair, India, Vice-Presidents: Dr. Francisco Mendieta-Jimenez, Mexico; Prof. Liu Jiyuan, China; Dr. Hiroki Matsuo, Japan; Prof. Anatoly Perminov, Russia, Secretary General: Dr. Jean-Michel Contant, France.

Activities:
Encourage international scientific cooperation through symposia and meetings in the area of: space sciences, space life sciences, space technology & system development, space systems operations & utilization, space policy, law & economy, space & society, culture & education; Publish cosmic studies dealing with a wide variety of topics including space exploration, space debris, small satellites, space traffic management, natural disaster, climate change, etc.

Cooperation with other Academies:
Establish cooperation with Royal Swedish Academy of Sciences (1985), Austrian Academy of Sciences (1986, 1993), Academy of Sciences of France (1988, 2001), Academy of Finland (1988), Indian Academy of Sciences (1990, 2007), Royal Spanish Academy of Sciences (1989), German Academy of Sciences (1990), Kingdom of Netherlands (1990), Academies of Arts, Humanities & Sciences of Canada (1991), U.S. Academy of Sciences (1992, 2002), U.S. Academy of Engineering (1992, 2002), Israel Academy of Sciences and Humanities (1994), Norwegian Academy of Science and Letters (1995), Chinese Academy of Sciences (1996, 2013), Academy of Sciences of Turin (1997), Australian Academy of Sciences (1998), Royal Netherlands Academy of Arts and Sciences (1999), Brazilian Academy of Sciences (2000), U.S. Institute of Medicine (2002), Academy of Sciences of Ukraine (2010, 2012), Academy of Sciences of South Africa (2011), Royal Society of South Africa (2011), Pontifical Academy of Sciences (2012).

Publications:
Publish the journal of the International Academy of Astronautics ACTA ASTRONAUTICA ranked 5th in the world (5000 pages per year); IAA Yearbook, Dictionaries and CD-ROM in 24 languages (last languages Afrikaner and Swahili); Book Series on small satellite, conference proceedings, remote sensing and history. All publications available at https://shop.iaaweb.org.

Membership:
Active members 1196 in 85 countries in four sections; Honorary members (3):
- Africa: Algeria, Burkina Faso, Cameroon, Egypt, Ethiopia, Ivory Coast, Kenya, Libya, Morocco, Nigeria, Senegal, South Africa, Tunisia.
- Americas: Argentina, Bolivia, Brazil, Canada, Chile, Columbia, Cuba, Ecuador, Guatemala, Mexico, Peru, USA, Venezuela.
- Asia: Azerbaijan, Bahrain, Burma, China, India, Indonesia, Irak, Israel, Japan, Kazakhstan, Kuwait, Kyrgyzstan, Malaysia, Mongolia, Pakistan, Saudi Arabia, Singapore, South Korea, Sri Lanka, Syria, Thailand, Turkey, United Arab Emirates, Vietnam.
- Europe: Armenia, Austria, Belarus, Belgium, Bulgaria, Croatia, Czech Rep., Denmark, Estonia, Finland, France, Germany, Greece, Hungary, Ireland, Italy, Latvia, Macedonia, Netherlands, Norway, Poland, Portugal, Romania, Russia, Serbia, Slovakia, Slovenia, Spain, Sweden, Switzerland, United Kingdom, Ukraine.
- Oceania: Australia, New Zealand.

IAA Office:
Office 6 rue Galilée, Po Box 1268-16, 75766 Paris Cedex 16, France; IAA Office Branches in Bangalore (India) and in Beijing (China) for its Study Center; Regional offices in Abuja (Nigeria), Tunis (Tunisia), Buea (Cameroon) and Nairobi (Kenya).

Mailing Address: IAA, PO Box 1268-16, F-75766 Paris Cedex 16, France
Telephone: 33 1 47 23 82 15, Fax: 33 1 47 23 82 16, email sgeneral@iaamail.org

Web Site: http://www.iaaweb.org
https://shop.iaaweb.org/